FEED THE SOIL

Edwin McLeod

Illustrated by:

Karen Blaesing   — plants & characters
Anne Alexander   — plants
Randy Griffis   — plants
Donna Landon   — calligraphy

Published by:

EDWIN J. MC LEOD
8855 KENBERTON
OAK PARK, MI 48237

To my family —
Genetic and Environmental —
To All.

Acknowledgements

Many people, through their support and effort, have helped to create this book. I would like to express my heartfelt thanks to my parents, my brother Skip, Joe Hines, Ben and Hazel Berg, Lloyd Andres, Kate Burroughs, Ron Whitehurst, Richard Applegate, charles chesney, Shery Litwin, Tom Roy, Kevin McSweeney, Amigo Bob Cantisano, David Salem, Bob Huffman, Robert Thomas and Lynn O'Neill. I also wish to thank all of my other friends for their encouragement along the way — the "How's the book doing"s. Your energy has helped too.

Peace,

Edwin

Contents

The Ride to Tomorrow

Rumble, bumble, bang, bam! "How the tracks had
deteriorated over the last decade," thought Hylas Hare.
He awoke suddenly from a dream – nearly fell out of his
seat, like most of the other passengers. That last bump
almost derailed the train. The passengers were on the
edge of their seats with anxiety. Would they make it to
Gitchee Gumee province?

Hylas, not the least bit worried or anxious about
the ride, resumed his daydream. A beautiful, young,
female rabbit with long black hair and dainty white paws
sat across the aisle from him. His mind wandered hither
and thither, wondering how he could strike up a conversation with her. You see, Hylas was quite shy. He would
ask her the time, but he obviously didn't need to know
since he had a watch on his wrist. And, after all, he
didn't wish to initiate an acquaintance through such a
trite matter, as time, or weather. It appeared that
Hylas would, as usual, not break the ice with this
attractive young lady. He settled for a romance in his
vivid imagination.

As Hylas dozed off again, a rather stout guinea pig
in the row ahead of them lit up a pipe. Great billows of
smoke wafted back into their seats. Almost immediately,
the dark haired girl began sneezing ferociously.
Without thinking – for, if he had, he would not have
acted – Hylas pulled his clean, white, monogrammed hand-
kerchief out of his front coat pocket.

She took the kerchief from Hylas, sneezing and
sneezing and sneezing. "This bunny must really be a
sensitive creature," he thought to himself. Her sneezing

fit would not cease, even though the guinea had consid-
erately put out his pipe. Hylas helped her to the next
car of the train.

They found a pair of empty seats and sat down
together. By now, her sneezing had died down to an
occasional outburst. Hylas introduced himself. The
young lady expressed her appreciation for Hylas' help.
She told him her name was Dahlia. Realizing he was in
the middle of a conversation with this beauty - sitting
right next to her - Hylas' heart began racing. Fortunately, his heart calmed after a few minutes of conversation, just as Dahlia's nose had.

"Where are you going?" asked Dahlia. "I'm off to a
summer homestead in Lucerne," replied Hylas. "I've
never been there before so I can tell you nothing more.
Luckily, I qualified as a seasonal farmer. I am on my
way to a parcel of land where I hope to farm indefinitely.
How about you Dahlia; where are you going?"

"I'm also going to Lucerne, Hylas. My parents used
to farm there. They farmed one hundred acres of land,
primarily covered with fruit trees. We also had a few
acres of vegetables which we sold, along with some of
the fruit, at a roadside stand. Unfortunately, my
parents couldn't make enough money on their produce to
pay the bills. Prices were too low and the market was
so bad that they had to sell the farm to a large corpor-
ation. But, that was a long time ago, nearly fifteen
years now. I was just in my early teens.

It was a trying time in my life. My poor father
loved the farm and worked hard to keep it. He and my
mother aged rapidly during those last, few years on the
farm. The entire experience wrung my heart. It left me
very remorseful. I am somewhat fearful in returning to
Lucerne because of those memories." Comfortingly, Hylas
patted her paw.

"It is funny how the tide turns," Dahlia went on.
"What do you mean?" queried Hylas. "For years, the
small farmers sold out to large farmers. The farms grew

larger and larger until only a few owners managed vast
tracts of land. These mega-farms were operated with
enormous inputs of fossil fuel - oil and natural gas.
You know the rest Hylas. The Fall is history now. How
strange life is Hylas. Here I am returning to my old
home town with a farmer."

Both seemed quite pleased with their future. It so
happened that this pair, like many other folks,were on
their way back to a life of farming. Quite simply, what
had happened was - the large land owners could no longer
farm profitably. The high prices of farm inputs and the
low prices they received for their crops put them out of
business. They stopped working and paying their taxes.
In fact, almost everyone did because the majority of the
economy was at a standstill. The government bought the
land from the large land owners. In order to put folks
back to work, they were redistributing the land for small
scale farming. This was an exciting time. There was new
hope for many who had spent four hard years of unemploy-
ment and hunger. Now, they could regain their self-
respect by growing food for themselves and others.

An agrarian renaissance was in the making. This
was catalyzed by two factors. First, and most importantly,
people had to be fed. Oil based mega-farms couldn't do
it any longer because two important resources, oil and
topsoil, had been depleted. Second, automation had
eliminated the majority of jobs in the factories and
offices so people were available for other livelihoods.
This set the stage for act one of a new agriculture.

"What is a seasonal farmer?" Dahlia asked. Hylas
thought for a moment. A seasonal farmer is a person who
lives and works in the city during the late fall and
winter months. Then, during the growing season, he lives
on, and farms, a two to twenty acre parcel. The seasonal
farmer actually owns his parcel. He signs a contract to
pay for the land over a period of time - in yearly in-
stallments, for a maximum of 20 years. I qualified to be
a seasonal farmer because I was a landscaper while I

attended college.

Bam! The train smacked another kink in the track, just before entering a tunnel. With this, Hylas' and Dahlia's paws met and clutched each others in fright. Their whiskers were pointing up toward the ceiling now; mouths wide open, revealing their big front teeth. They looked as though they were cringing with the expectation of having their teeth drilled by a dentist.

When the train came out of the tunnel, the conductor shouted, "Next stop, Lucerne!". Innerly embarassed that they found themselves holding each others hand, Dahlia exclaimed, "I hope my grandfather made it to the train depot alright. He is getting up there in age; almost 85 now, but he still works just as hard as a middle aged rabbit. I'm going to live with him. I trust that Lucerne will bring me as long and healthy a life as my grandfather." With this last sentence, Dahlia saw her grandfather waiting on the platform. The train came to a stop. Dahlia jumped out of her seat, quickly shook Hylas' hand and said goodbye. It was clear that she was excited to see her grandfather. Dahlia was the first one off of the train.

Hylas was dumbfounded by her rapid departure. When his senses returned, he thought to himself in dismay, "Why, I didn't even get her address. What a slow mover I am - what a dummy - the most beautiful woman I've met in years." Hylas continued to lambaste himself as he slowly made his way off of the train. But, behind all this, he was glad to arrive in one piece.

৺ Agricultural Renaissance ৶

Hylas made the long step off of the train. Actually, he jumped. He wasn't carrying very much; just some clothing, sleeping bag, a little food and his seasonal farmer qualifying letter. The first thing that caught his eye was a large sign with green lettering which read "Welcome Seasonal Farmers. Assemble Here.". A small crowd of rabbits and chickens were standing together sniffing and cackling; causing a great stir. Hylas slowly approached the crowd. He didn't recognize anyone.

Just as he came within view of a man on a soapbox, the elevated gentleman spoke, "Alright, quiet down everyone. Let me have your attention for a moment. I have some good news and some bad news for you seasonal farmers. First, the bad news: We don't have any plush hotel with showers and heat for you to sleep in tonight. I'd tell you to sleep under the stars, but it seems there won't be any stars tonight. We are expecting a thunderstorm. Therefore, you will all have to huddle together like bunnies under the awning of this platform." A great moan came from the audience. "Now, if you all are considerate of each other, we won't have any trouble. Everyone will get a good nights sleep. Let me warn you. You better get a good sleep because tomorrow we will be hiking some long distances to issue each of you your land for farming." A cheer came up from the crowd.

"Now for the good news: We don't have fancy sleeping quarters for you tonight, but we do have a fancy meal. If you all will kindly proceed over to the north side of the depot, you will find a trough where you can get washed up. Hot meals will be passed out the kitchen window next to the trough."

Next morning, the sun was shining brightly in a blue
sky. The ground was soggy and puddled. Everyone was in
good spirits, even those who had slept close to the edge
of the platform and were a little damp. They knew that
the sun would quickly dry their clothing. Hylas was one
of the more fortunate creatures. He was perfectly dry
because he had gone to sleep early and found a sleeping
spot well under the platform. Those who stayed awake,
singing, playing instruments, talking and watching the
weather, ended up with the less sheltered spots.

Hylas was doubly lucky because he had gotten plenty
of sleep. You see, the soap box rabbit rang a bell just
before sunrise. He announced, "Everyone up! We will be
leaving at sunrise." Then he proceeded to tack up four
signs. They read: NORTH - no.s 1-100, SOUTH - no.s 101-
200, WEST - no.s 201-300, and EAST - 301-400. "Find
your number on your qualifying letter," he shouted. The
group of drowsy rabbits and chickens slowly arose, like
wilted plants after being watered. Each one pulled out
his letter. They began assembling around their appropriate sign. Hylas was number 10.

The North sign had about two dozen seasonal farmers
gathered around it. After a few minutes had passed, the
soap box rabbit approached and said, "Hello again, let me
introduce myself properly this time. My name is Jahmay.
I will meet each of you along the way or as I issue you
your land. Try to meet as many fellows as possible
while we hike. After all, they will be your future
neighbors. Follow me folks." With that the group of
them headed north, out of the town of Lucerne.

It was only a short distance to get out of town.
Lucerne was quite small. They hiked over an old road,
passing through a large meadow dotted with trees. The
trees were mostly large old oaks, poplars, willows and a
scattering of 3 to 4 year old saplings - indicating the
length of time which had passed since the land had been
cultivated. Most of the fields they passed were covered

with last years dried weeds and small pools of water.
Everyone in the group was pleased to be hiking on such a
beautiful morning.

Jahmay and Hylas now walked together. As they passed
an old dairy farm in good repair, Hylas asked Jahmay,
"Who is farming that land?". "Why, a family of cows own
that dairy farm. Haven't they kept it up well? The cows
were able to hold onto their small 40 acre place so now
they continue to operate their family dairy.

"You will eventually discover that there are quite a
few family farms in operation, mostly dairies. In fact,
many cows are returning to the country to start family
farms. You see, the cows are best suited for family
farms because they can produce throughout the year.
They can keep producing milk here in the country during
the winter, while you return to the city for your work.
You may think that the cows have it good, being able to
stay in this rural paradise throughout the year. Really,
each farming opportunity has its own special advantages.

"Let's take the gypsy farmer, for example. Maybe,
you haven't heard of this type of farmer yet so I will
explain. Sheep qualify to be gypsy farmers. They work
under the management of a sheepdog. The sheepdog does
all of the planning and coordinates the activities of
the gypsy farmers - what will be planted and when to weed,
for example. He has a big responsibility. Fortunately,
managing sheep is what the sheepdog enjoys most so this
works out fine.

"The advantages of being a gypsy farmer are: you
don't have to take any responsibility and you can travel
from gypsy farm to gypsy farm, experiencing different
crops and parts of the country. Of course, if a gypsy
farmer wants to settle down, he can take farm classes
while gypsy farming. This will enable him to qualify as
another type of farmer." Hylas nodded with understanding
as Jahmay rambled on. Jahmay loved to hear himself talk.
But, why not? He spoke very well.

Hardly pausing for a breath, Jahmay said, "Here we

are at our first stop. Alright, the first four seasonal
farmers on the list are: Glabella, Polaris, Epictetus
and Origen. As you can see, your parcel has your number
affixed to it, hanging on a post. A survey team has
plotted out your land and a reforestation group has
planted a dwarf tree at each corner of your property.
There are instruction sheets attached to your number.
In this packet, you will find directions to the supply
house and seasonal farmer community center. What I
want you four to do now is look over your land and think
for a few hours. Think about what you are going to grow
here. Introduce yourself. You are a newcomer to this
place."

They ventured on. Hylas asked Jahmay, "Who is
farming the land surrounding the first four seasonal
farmers?" "Well," Jahmay started. He had to finish a
thought before answering this question. "Most of the
surrounding land will be cultivated by gypsy farmers.
The rest will be replanted with native trees by the
reforestation service. We want to develop a balance
between farmland and natural reserves. The natural
reserves will provide a home for wildlife, as well as
give farmers a place to enjoy on their days off."
"I see," replied Hylas. Jahmay continued, "I suspect
that as people learn more about farming there will be
more seasonal farmers like yourself. The gypsy farmers
will claim stewardship of their own parcels."

"Hey, look at the skunk cabbage!" someone shouted.
All of the rabbits' ears perked up when they heard
cabbage. A few of them even began walking off the road
in order to examine this purplish plant, but they had to
turn back because the soil was much too soggy. The
sight of the skunk cabbage lightened everyones' spirit,
having seen the first flower of the year on such a
pleasant day.

They soon came to the next set of four parcels.
After issuing the land, Jahmay gave the same, small
speech as before. One of the parcels belonged to Hylas

- a five acre plot. His heart jumped into his throat as
he walked onto his soil. This land was under his care
now, as much as his own body was. He was very conscien-
tious when it came to clipping his nails, combing his
hair, washing, brushing his teeth, eating properly -
never too much, drinking plenty of water and exercising
daily. He didn't expect any problem being a steward of
the land. But, where to start?

As if in a dream, Hylas strolled around his five
acres. A flock of geese flew overhead. They all waved
hello. Hylas waved back. As he was looking up to watch
the geese on their northward journey, he stumbled into a
thistle patch. "Owwwww!" he screamed. "What a way to
get acquainted with my new summer home. You will be
the first to go." he thought. But, these were not
ordinary thistles. They were giants - almost six feet
tall. "Well," Hylas said to himself, "if this place
grows such huge thistles, I'll bet it will grow large,
tasty vegetables - why, carrots even." With this sentence, he pushed back some of the dried weeds and dug
down into the soil. It was a loose, clay loam. He
could dig down without much trouble, except for the
abundance of weed roots in the way. This pleased
Hylas. He even forgot about his thistle wounds.

After completing a circular walk around his property,
Hylas returned to his numbered post. The instruction
pamphlet practically jumped out and bit his brain. He
had forgotten all about it. Flipping open the first
page, he saw a map to the store and community center.
"Oh no," he said. "I was supposed to go pick up my tools.
How long have I been dilly dallying here?" Off he
hopped.

It was fortunate that Hylas had not forgotten to
bring the map and had hopped all the way to the center.
The clerks were about to close the store. Any other day,
the clerks would have been upset with a customer arriving
just before closing time. After all, they had worked a
long, hard day and were tired. But, today was different.

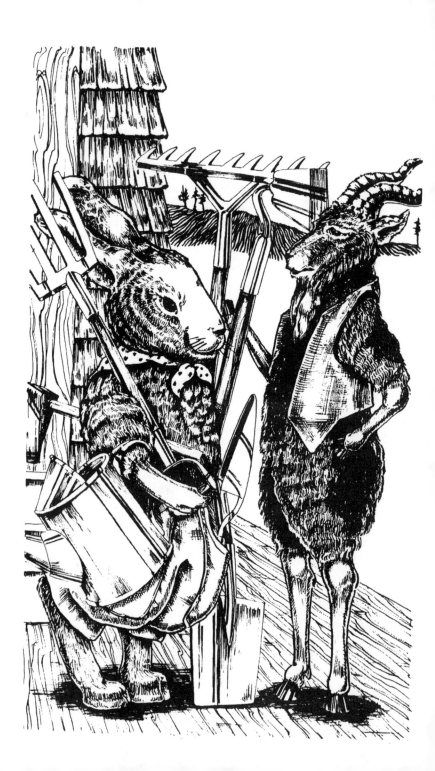

It was so beautiful outside that everyone was in good spirits. Because the clerks had been waiting on the newly arrived, seasonal farmers all day, they knew exactly what Hylas needed. Like dancers, they pirouetted from shelf to shelf, rack to rack, assembling Hylas' farming equipment.

"Let me see," Hylas pulled out his instruction phamplet. "I need a hoe, spade, trowel, rake, scythe, fork, hatchet,watering can, broadcast seeder and wheel barrow." A shaggy headed clerk brought a trowel. As he set it in the pile with the other tools, he said, "We have everything you need, except a wheel barrow. Come back tomorrow afternoon and we may have it by then. We are expecting a shipment of tools in the morning."

"Oh well, it could have been worse," replied Hylas. "At least I have everything else." He began picking up his tools. The clerk helped him get loaded. On his way out the door, a goat stepped in. "Well, hello rabbit. I saw you hopping so fast earlier that I thought a fox was chasing you." The goat was a very friendly fellow. "So this is where you were rushing to." "Sorry, we've just closed, Gotaway," said the clerk. "Well, that's alright. I'll just come back tomorrow morning," Gotaway sang politely.

"Now you know my name. What's yours, rabbit?" "Hylas," he muttered. The weight and awkwardness of all the tools Hylas was carrying was becoming a great strain. The watering can was about to slip out of his grasp. He wished this friendly goat would let him pass and skip the small talk. "Here, let me help you with your load Hylas. I'm headed back in the same direction as you. I'd be happy to lend you some assistance." Gotaway grabbed the watering can and half of the tools. Out the door they clomped. Hylas was relieved.

"I'm fortunate to have your help, Gotaway. Thank you. If my wheel barrow would have been there, with the rest of my tools, I wouldn't have had any trouble." Gotaway stopped to scratch his head on an old tree.

"Cooperation, that's my middle name," replied the goat. "Why, cooperation and hard work is what makes life worth living. Or, I should say, it keeps us living, or . . . you know what I mean. We all live for the mutual benefit of one another. Don't you think so Hylas?" "Yes, I think so," answered Hylas. "But, I haven't thought about it much."

   "I can't help but realize it every day," said Gotaway. "I live on a collective farm. You passed it in your hurry; probably didn't even see us building that new dairy barn." He pointed to a new structure off in the distance. "We do almost everything in a group on our farm. There are twenty five members in our collective. Of course, there are advantages and disadvantages to belonging to a collective, but for me, there are definitely more advantages. Heck, I love to be with my fellow goats. In a herd of twenty five goats, I know that we always have the power to accomplish almost any task. Even more important, if I ever want to talk to anyone, I know there is always someone who will swim in the breeze with me.

   "The biggest problem on our collective farm is in decision making. Seems, we have too many meetings and waste too much time in discussion. I'm sure we will gradually organize ourselves so there will be more individual decisions and fewer group decisions. This way, we won't need to have so many meetings. Meetings will be required only for the really major issues.

   "Let me be frank with you Hylas. Our collective has its skirmishes now and then. What goats don't? - especially during rutting season. We're cooperating to minimize our skirmishes. In fact, last month we initiated a plan of critique days. One week before a member's birthday, every member of the collective writes a commentary on his or her impression of the birthday person - his work and relationship with others in general. This really helps the individual grow and the collective release its feelings. Everyone knows how it feels to be the recipient of a critique so the members do it with compassion. By the time the birthday arrives, everyone feels lighter about their relationship with the birthday person. The celebration truly becomes a rebirth."

During their conversation, they passed the new dairy of the goat collective. Gotaway explained all the important features of the new building and their milk operation. Like the cow family farm, the goats stayed on their dairy farm all year. Hylas was intrigued by the collective, although he thought operating his own farm would suit him best. They were now approaching Hylas' place. Gotaway had kindly carried half of the load past his own farm. Hylas was very grateful. After dropping the tools, they said goodbye. Gotaway quipped as he left, "Next time, you'll tell me all about yourself."

Who Really Feeds the Plants

The following morning was just as beautiful as the previous day. A few clouds, high in the atmosphere, heralded the beginning of a warm front, promising rain the next day. The birds were singing merrily as Hylas walked home from the community center. He had just finished breakfast with his fellow seasonal farmers. Hylas could have eaten dinner at the community center on the previous evening. However, he didn't know what the eating arrangement was, having not read his pamphlet thoroughly. Anyway, he was so tired after carrying the tools home that he went to sleep immediately.

Now, he was glad he had gone to sleep early because he was energized and ready to work. "Well, I might as well get my spade and start," he thought. Hylas pierced the soil with the spade and out bellowed a scream, "Eeeeeooow!". Out of the soil squirmed an earthworm. "Oh no, don't tell me you farmers are back. Oh woe is me. Woe is my family." The earthworm started to weep. "What am I going to do. I have just about completed rebuilding my home. Once again, I can afford to feed my children properly. Now you're back. Oh, oh, ah, hah, hah!" The earthworm began moaning and bawling even louder than before.

"Wait! Stop crying earthworm. You must have mistaken me for someone else. I have never done anything to hurt you before. Why, I've never seen you till now," exclaimed the rabbit.

"You're right. You haven't. But, all of you farmers are alike," said the earthworm. "You murder us with your pesticides and fertilizers. You destroy our

homes and food by <u>not</u> allowing the plants to decompose
in the soil as they properly should. Why, you, you,
you . . . are a pestilence on us earthworms and other
soil creatures," he stammered. The earthworm went on
sobbing.

Hylas became very distraught. "Mr. Earthworm, sir,
I'm different. At least, I hope to be . . . I mean . . .
a better farmer than the others. What can I do to be a
better farmer, Mr. Earthworm? How can I help you? Can't
we work toward each others mutual benefit — you know,
cooperate. Who are these other soil creatures? I've
never met them before. Heck, I never even knew anybody
lived down there, except a few worms, like yourself, and
maybe some ants, beetles and other bugs. The soil was
just soil. You know, like water is water."

"Ah! That's right. You're just like the other
farmers. I was right. They didn't even know we existed."
The earthworm screamed in sorrow. "You're just like
them." "No I'm not," Hylas pleaded. "Now I know you
exist, Mr. Earthworm. Sure, I don't know as much as I
should, but I'm certainly willing to listen and learn
whatever I can about the soil and its inhabitants."

After a while, the earthworm calmed down. "Well,
maybe you are different. That's true. The others never
did even stop to speak with me; as if I didn't even have
a voice or something, or like they were deaf. Sit down
then. What's your name?" "Hylas," the rabbit replied.
"Mine is Eatmore. Nice to meet you. And please, don't
call me Mr. Earthworm anymore. I can't stand to be called that. Now, make yourself comfortable Hylas. Relax.
I have a great deal to tell you. This is the most important thing I'll ever say: <u>Feed the soil organic matter —
dead plants</u>. Wait, you don't even know who you are
feeding. I know, I'll start by introducing you to my
friends; even though you won't be able to see most of
them because they are so small."

I'll introduce them from smallest to largest:

Bacteria

The bacteria are single-celled creatures which live alone or in groups. They are so small that a lump of soil, the size of a pea, may have up to a billion of them. This is not surprising because bacteria reproduce very rapidly. They split apart to form new cells. In the morning, there may only be one bacterial cell, but by the afternoon this one may have split into billions of cells - provided they have food and the right growing conditions. If conditions become unfavorable for bacteria, many species can produce spores which are resistant to a harsh environment. This characteristic is quite helpful because bacteria generally are less mobile than other organisms and cannot escape from unpleasant situations.

Most soil bacteria are found in the first foot of topsoil. They are probably the most important soil microorganisms because they can eat nearly anything, except large, complex molecules. Bacteria break down organic matter into humus and nutrients for themselves and plants. They digest their food by secreting enzymes to dissolve organic materials. Think of it like bacterial saliva. The food that is released diffuses to them.

Bacteria come in many shapes - primarily rods, spheres and spirals. The species which have more surface area, like the spiral shaped ones, can absorb nutrients most efficiently. In lakes and oceans, where the concentration of nutrients is low, most of the bacteria are spiral or rod shaped. In nutrient rich places, like some soils, the sphere shaped bacteria are more common. An abundance of nutrients in soil is not the only reason why spherical bacteria are more common there. Sphere shaped bacteria are usually more resistant to drying than elongated ones so they can grow outside of the moist lakes and oceans.

Remember Hylas, most soil bacteria are your friends. They aren't fussy about what you feed them, but make sure you feed them organic matter. /j1, m17/

Actinomycetes

Actinomycetes are soil molds. They produce an extensive, threadlike network within the soil. Bacteria generally work within the top 12 inches of soil. Actinomycetes, on the other hand, may work many feet below the
surface. These soil molds are important decomposers of
organic matter and can be very prolific. They comprise
about 20 percent of the microorganisms in an average
soil. Soils high in organic matter may contain 40 percent actinomycetes. When large numbers of actinomycetes
are working, you can smell a musty, earthy odor rising
from the soil - sort of like a rotting log.

Actinomycetes can be rather antisocial at times.
When they don't want to be crowded by rapidly reproducing bacteria they produce antibiotics which inhibit
the growth of bacteria. Very acid soils inhibit the
growth of both, bacteria and actinomycetes.

These soil molds eat in the same manner as bacteria,
although they can eat much larger, complex molecules.
They secrete digestive enzymes which break down two of the
most complex molecules in organic matter, namely:
protein and cellulose. Actinomycetes are also remarkable
because they are efficient users of nitrogen. For every
50 parts of a carbon source (like sugar or cellulose)
which they break down, soil molds need only one part of
nitrogen. In comparison, bacteria can only break down
20 to 30 parts of carbon for each part of nitrogen. I
will explain this advantage in more detail later on.
/r4, m17/

Fungi

Fungi are primitive plants. They do not have chlor-
ophyll (the green stuff), like other plants, so they
cannot produce their sugary food through photosynthesis.
Alternatively, fungi obtain their food by breaking down
organic matter.

Most fungi need oxygen to eat and grow. However,
they aren't as particular about their growing conditions
as bacteria. Fungi tolerate a wider range of pH than

bacteria - 5.5 to 8.0, as compared to 6.0 to 7.5 for
bacteria. Additionally, fungi are more efficient at
breaking down carbon sources than bacteria so they don't
need as much food. You can see that fungi fill an important role in the soil community. /a6, b5, g5, m17/

Earthworms

Now, I'll tell you about myself. Aristotle once
said we were the intestines of the soil. He was right.
I consist mainly of one long intestine. Throughout my
lifetime, I eat organic matter and deposit earthworm
manure (castings) in and on the soil. These castings are
the richest and finest quality of all soil material.
Fresh castings are very high in bacteria, available
nitrogen, calcium, magnesium, phosphorus and potassium.
An average earthworm, like myself, can produce its weight
in castings in 24 hours. As much as 15 tons of dry soil
can pass through my body in one year. Not bad, eh?
/r4, m17/

I don't mean to boast Hylas, but earthworms are the
world's premier cultivators of the soil. We burrow into
the soil, aerate and mix it; sometimes as far as six feet
deep. Do you have a plow or rototiller that will go
that deep?

You may not have noticed yet, my Boy, but I am
neither a male or female, but both. Yes, I have both
sexes. You may think this is complicated. I find it to
be fun.

We're such good guys - or should I say gals? -
wouldn't you like to have more of us around? Well Hylas,
under favorable conditions, I will produce an egg capsule
every week to ten days. You can make the environment
more favorable for earthworms by adding more organic
matter, cultivating as little as possible, providing
enough lime and moisture, and avoiding the use of pesticides and chemical fertilizers. Two to three weeks after
the egg capsule is deposited, the wormlets emerge. One
mature worm will beget over 150 worms each year of its
life. The average lifespan of an earthworm is two years.

I'm not going to be around long Hylas so I hope you are listening closely. /e1, r4/

One area earthworms can't take all the credit, is in our work as physical decomposers of organic matter. What I mean is: breaking large chunks of organic material into smaller sized pieces for bacteria, actinomycetes and fungi - so they can eat them faster. Millipedes, sowbugs and spring tails can also take some of the credit. Did you see them take a bow? I might also mention in a whisper the mites, centipedes, spiders, beetles and some other - very quietly - predators. You know, I don't want to attract the predators attention. If I did, my decomposing friends would be extremely upset with me. They get harassed enough as it is, without my calling for their enemies. /m17/

This leads me to what all of us soil organisms are working toward - Optimum Plant Growth. You see Hylas, the better the plants grow, the better we are fed because there are more dead plants (organic matter) for us to eat. This is true, at least, when you so called farmers are not stealing our harvest. We soil organisms are truely the world's first and oldest farmers. We have been growing plants long before you appeared on the scene.

Are you listening Hylas?" "Why, yes Eatmore," he replied. "You shouldn't need to ask. Can't you tell I'm paying close attention?" "Sure, it appears that you are, but I'm not guaranteed that it is sinking in," said Eatmore. "Let's take a break Hylas. Give my mouth, your ears and our brains a rest. We can continue this lesson with renewed energy in 15 minutes. I am going to tell you about the growing needs of plants next. We should both be alert for our communication to be effective." With this, Eatmore slithered back into his cool hole.

~ Soil Diet —
Plant Health ~

For optimum growth, plants need: 1. water, 2. nutrients — both macro and micro, 3. good, loose soil
structure, 4. sunlight, 5. proper temperature conditions,
6. adequate humidity, 7. clean air and 8. not too much
wind. All of these requirements are important and all
are interrelated. You could try to optimize one requirement, but, in reality, they must all be considered together. For example, let's say you gave your plants just
the right amount of water, not too much and not too
little, but the spot you chose was too windy. Your
plants were blown over. Hylas, all of the growing needs
of plants must be met, or no growth. You must consider
all of them together. I could say the same thing about
your growth or mine. I'm sure this is easy for you to
understand.

The requirements which we can influence most readily
are the soil factors: water, nutrients and structure.
How do we do this? Well, the process revolves around
organic matter — the food of soil organisms.

Nutrients *

Macro	Micro (some)
Nitrogen	Boron
Calcium	Iron
Magnesium	Manganese
Potassium	Zinc
Phosphorus	Copper
Sulfur	Molybdenum
	Chlorine

* Micro needed only in trace quantities.

When plants die and drop to the ground, or trees lose their leaves in the fall, they turn into organic matter. Bacteria, actinomycetes, fungi and earthworms eat organic matter. It may seem funny to you Hylas, but we soil creatures need the dead bodies of plants to feed ourselves. Later on, we return the favor by feeding the plants. This is the grand cycle.

Organic matter consists of large molecules like protein, cellulose and hemicellulose (tough, fibrous or woody material), lignin (the glue which holds the fibrous material together), fats and waxes; and some smaller molecules like sugars and starch. The main element which these molecules are composed of is carbon. The carbon content of organic matter is between 40 and 60 percent. The nitrogen content, on the other hand, is considerably less – between 3 and 6 percent. Nitrogen is contained primarily in plant protein. As plants get older, the percentage of nitrogen decreases. /b5, f3, v1/

The Main Elements of Organic Matter		
Carbon	40 – 60	%
Oxygen	30 – 40	%
Hydrogen	3 – 6	%
Nitrogen	3 – 6	%
Sulfur	2 – 5	%

We decomposing soil organisms eat organic matter – the protein, cellulose and hemicellulose, lignin, fats, waxes, sugars and starch, breaking them down into smaller, simpler molecules. These smaller molecules are easier for the plants to eat. We help the plants grow by pre-digesting their food.

We pretty much need the same growing conditions as you rabbits. Except, some soil organisms do not need oxygen. These organisms are called anaerobic decomposers. The soil organisms which decompose organic matter faster and do more to improve soil fertility, do use oxygen.

They are aerobic decomposers. Soil organisms which
decompose organic matter aerobically are more desirable
because they create soil conditions beneficial to plant
growth. Anaerobic decomposition produces substances,
such as alcohols and ketones,which inhibit plant growth.
This type of decomposition is similar to fermentation -
like when you make wine. Plants can't tolerate much
alcohol. Aerobic decomposition is preferred by plants.

The aerobic soil organisms need plenty of oxygen.
That's why good, loose soil structure is important - so
oxygen can penetrate the soil easily. Organic matter
loosens and fluffs up the soil, helping oxygen get to the
aerobic decomposing soil organisms.

The rate of aerobic decomposition increases as the
availability of oxygen increases. Too much water can
hinder the movement of oxygen into the soil. For example,
oxygen cannot penetrate the soil when water has formed
puddles over it. Organic matter helps prevent water
from standing in puddles. It acts like a sponge to
soak up water, leaving passages for oxygen to penetrate
into the soil.

Aerobic soil organisms do not like to be flooded
out. We function best with just the right amount of
moisture. Our fastest rate of digestion is at 60 to
80 percent of the soil's water holding capacity. However, it is better if the soil is not moist continuously.
We eat faster when the soil is exposed to a cycle of
drying and wetting. Several of these cycles stimulate
our digestive activity. /a6/

Another reason why we eat slower in soils that are
too wet is because they are too acid, or, in other words,
have a low pH value. The "pH value" is a measurement
of the quantity of free hydrogen ions present. The
greater the quantity of hydrogen ions, the lower the
pH value. For example, a soil with a pH of 4.5 is very
acid, or has a large quantity of hydrogen ions, while 6.5
is slightly acid. A pH of 7 is neutral. The pH values
larger than 7 are alkaline. Decomposition is most rapid

in neutral to alkaline soils. Adding lime (Calcium carbonate) generally makes a soil more alkaline so it speeds up microbial activity and decomposition. Some folks give soil pH a great deal of attention, but there are other factors, like temperature, which are just as important. /a6/

Believe it or not, we soil organisms do quite a bit of eating and growing at soil temperatures of 40°F (5°C) and sometimes lower. As you would suspect, our rate of digestion increases as the temperature increases; up to 85°F (30°C), where the rate stays about the same. Fortunately for me, it rarely gets that hot in the soil. I like a change of weather Hylas, but not the extreme heat. When the soil temperature rises too high, I burrow deep into the cooler soil. /a6/

One thing that really slows us down is not having enough nitrogen in our organic matter. This deficiency slows down plant growth too. If there is not enough nitrogen in our food, organic matter, we cannot continue to eat, grow and reproduce. There really isn't an exact amount of nitrogen that we and the plants need. It varies according to the quantity of carbon in the organic matter. The more carbon in organic matter, the more nitrogen we need in order to digest the carbon.

There is usually plenty of carbon available for us to build our bodies. Unfortunately, there is often not enough nitrogen. The quantity of carbon compared to nitrogen is expressed in a ratio - the carbon to nitrogen ratio (C/N). The optimum carbon to nitrogen ratio for rapid decomposition of organic matter is 15 - 25 parts of carbon to 1 part of nitrogen (15/1 - 25/1). When the nitrogen content of organic matter is low, the C/N ratio gets wider, for example 40/1. This ratio varies a great deal for different sources of organic matter. Decomposition slows down as the C/N ratio widens. You can speed up decomposition by adding nitrogen to organic matter. Some natural sources of nitrogen are chicken manure, fish emulsion or blood meal. /f3, g5, o1/

C/N Ratios

Young sweet clover	12/1
Alfalfa hay	13/1
Rotted manure	20/1
Clover residue	23/1
Green rye	36/1
Corn stalks	60/1
Oat straw	74/1
Timothy	80/1
Sawdust	400/1

/r4/

If there is little nitrogen in organic matter and the C/N ratio is wide, we soil organisms use up this nitrogen in building our bodies before the plants get to it. There isn't enough nitrogen to go around for everyone so the plants get deprived of their share of nitrogen. Their growth slows down and they become yellow. Acquaint yourself with the C/N ratio of different sources of organic matter so you do not shortchange us or the plants. Legumes generally have more nitrogen in them and have a narrower C/N ratio than grasses or other non-legumes.

Something else important to remember is that small, simple forms of nitrogen are very water soluble. They can be washed away easily. Large molecules with nitrogen in them, like organic matter, are not as water soluble. They hold and store nitrogen. As organic matter is digested by soil organisms and they die and decompose, nitrogen is released and becomes available to plants. This is sort of a time release system. It is good to have plenty of organic matter with the optimum C/N ratio in the soil. Soils which have enormous quantities of organic matter in them, like peat, have large reserves of nitrogen. The larger the percentage of organic matter in a soil, the larger the nitrogen reserve will be. For example, a soil with 2 percent organic matter can have an organic nitrogen reserve of 2,000 lb.s/acre and a soil with 4 percent can have 4,000 lb.s/acre. /w6/

As decomposition progresses and we digest more and more organic matter, the carbon to nitrogen ratio narrows. For example, it may change from 30/1 to 15/1. The reason this occurs is because after we have eaten and digested organic matter, we excrete carbon dioxide. Carbon is lost. It goes up into the atmosphere as a gas, carbon dioxide, or carbon is washed away as carbonic acid (the stuff in carbonated soft drinks). Nitrogen, on the other hand, continues to cycle in the bodies of soil organisms and plants when the soil is alive. More and more carbon is lost from the soil until the carbon to nitrogen ratio narrows to around 10/1, the C/N ratio of soil microorganisms. /f3/

As decomposition progresses and the C/N ratio narrows, the soil changes through a succession of different species of decomposing organisms. The first successional group of decomposing organisms love to eat the sugars and starch. This is not an unusual taste preference, eh Hylas? Now remember, sugar and starch are the simpler, smaller molecules in organic matter. They are so simple and small that a large number of bacterial species can eat them.

The second successional group of decomposing organisms can digest the larger molecules. The easiest molecules for them to digest are first, the proteins, then celluloses and hemicelluloses, lignins, fats and waxes. The secondary group also eat the primary group of decomposers themselves and their exudates. Then a third group will arrive to continue eating the large molecules and the dead bodies of the second and first group - a pyramid of decomposers results. As the decomposers higher up on the pyramid eat the lower decomposers, more and more carbon dioxide is given off until the C/N ratio approaches 10/1. /a6, g5/

Organic matter is digested by the successional stages of decomposing soil organisms until it cannot be broken down any further. The material which cannot be broken down any further. The material which cannot be digested is left behind as humus. Humus is very important for soil fertility and plant growth so listen closely Hylas. It is important because humus acts as a magnet, holding positively charged nutrients until the plants can use them.

Humus is a dark substance composed of many, many super enormous molecules which are primarily made of carbon. This dark soil substance is not understood very well because it is made of so many different large molecules. What we do know is that there are three groups of large molecules which are constituents of humus. They are: fulvic acid - which is soluble in acid and alkali, humic acid - soluble in alkali, and humin - soluble in neither acid nor alkali.

Now, I said that humus acts like a magnet. Let me explain what I mean.

There are some soil nutrients that are mobile and others which are immobile. The mobile nutrients readily dissolve in water and go where water goes. The immobile nutrients do not readily dissolve in water, but are, instead, held by humus and clay particles.

Nutrients

Mobile	Immobile
Nitrogen	Phosphorus
Sulfur	Potassium
	Magnesium
	Calcium

/b4/

The mobile nutrients, nitrogen and sulfur, are held within organic matter and the bodies of soil organisms. When the soil organisms die and break down, these mobile nutrients become dissolved in the soil water, or, also known as, the soil solution. The soil solution comes into contact with plant roots and the mobile nutrients are taken up by the plant. These mobile nutrients can also be absorbed by soil organisms or washed away and lost. As the soil dries out, the soil solution becomes more concentrated with mobile nutrient ions. The diffusion rate toward the roots is faster as the concentration of nutrient ions in the soil solution becomes higher.

Plants must be supplied with adequate nutrients during their entire lifetime. They need a slow release of mobile nutrients, like nitrogen, for optimum growth. Nitrogen and sulfur are required by plants in large amounts. At any one time, they are present in the soil solution in relatively small concentrations. Organic matter and the bodies of soil organisms act as millions of small time release capsules. As they decompose, nitrogen and sulfur are slowly released. The plants gradually consume these two mobile nutrients. Until organic matter and soil organisms decompose, the mobile nutrients are held in reserve for later use by the plants. The quantity of mobile nutrients in a soil depends largely on its organic matter content. /b4, f3, m13/

Plants can be given too much of a nutrient. For example, when soils have been fertilized with too much chemical nitrogen, nitrate can accumulate in plants at levels which are toxic to animals and humans. Additionally, when too much chemical nitrogen fertilizer is applied to the soil, this mobile nutrient is easily washed into our water system - ground water, rivers, lakes and oceans, acting as a source of pollution. /r5/

Fortunately, on heavy textured soils, like clays, nitrate will not leach rapidly. Instead, nitrate accumulates in the subsoil. Crops with extensive root systems, such as grains, may absorb subsoil nitrate. This nitrate may be adequate for crop growth. In some cases, it may even be too much, causing lodging (falling over) of grains. This situation most frequently occurs in the Great Plains states, Washington, Montana and North Dakota. Testing the subsoil helps you determine whether an adequate quantity,or too much nitrate is present in the subsoil. /b4/

Likewise, in soils which have been excessively fertilized with phosphate, large deposits of phosphate have
been building up in the subsoil. They get precipitated
on minerals like aluminum and iron. These precipitated
phosphates can be utilized with biological activation
of the soil - more soil life. In biologically active
soils, phosphate becomes dispersed in living and dead
organisms and organic matter. Acids from the decomposition of organic matter, like carbonic acid, liberate
phosphates from the minerals which they are precipitated
on. Without organic matter and bio-activation, phosphate
is only available from the soil solution. In live soils,
rich in organic matter, phosphate is continually being
freed from the subsoil minerals by acids created through
decomposition. The roots of plants take up this phosphate
as it is released. Phosphate is also gradually fed to
the plants as organic matter breaks down and soil organisms die. However, this topsoil phosphate does not have
to be taken up by plants immediately. It can be held by
the clay-humus magnet; ready for the plant to absorb it
when the roots come in contact with it. /w6/

Clay and humus bond together in soil. Together or
separately, they act as magnets to hold immobile nutrients
- potassium, magnesium, calcium and phosphate. These
immobile nutrients are positively charged. They are
called cations. The outer surface of clay and humus particles - our magnet - is negatively charged.

You may have heard about the Cation Exchange Capacity
(CEC) of a soil, Hylas, and not understood what it was.
Cation Exchange Capacity is a measurement of how strong
your soil's clay-humus magnet is, or, in other words,
the quantity of cations your magnet can hold. This
figure goes up to 50. Good compost with plenty of humus
has a CEC in the 40s.

The clay-humus magnet in soil holds cations until
plant roots come in contact with them. The plant roots
must forage these nutrients. Therefore, the volume of
soil which the roots occupy is crucial in the plant's

ability to take up immobile nutrients. Once again, we can see how important loose soil structure is - plant roots can spread out more readily in loose soil. They must come in direct contact with the clay-humus magnet in order to obtain a cation. The plant root takes up the cation and leaves a hydrogen ion in its place.

Some cations are held more tightly to the clay-humus magnet than others. The cations which are not held so tightly are more readily exchanged with plant roots. If an easily exchanged cation is present on the clay-humus magnet, it will be much more readily available to plants than a cation which is tightly held. Consequently, it is just as important to know how a cation stands in relation to other cations - which one is most easily exchanged and which is held tightest - as how large a quantity of each cation there is in the soil.

	Cations
Easily exchanged	Sodium, Ammonium
	Potassium
↓	Hydrogen ion
	Magnesium
Held most tightly	Calcium

Let me give you an example. Potassium is more easily exchanged than magnesium. Therefore, if there is more potassium on the clay-humus magnet than magnesium, potassium can inhibit magnesium release. A large amount of potassium in a soil may provoke a magnesium deficiency in plants, even when the magnesium level appears to be sufficient. Magnesium deficiencies are the cause of grass tetany in cattle. Similarly, sodium and ammonium are known to inhibit the release of potassium, magnesium and calcium. Even hydrogen ions in too acid a soil will depress magnesium and calcium release.

Due to the difference in the ease of exchangability of cations, you want your soil to contain more of the

cations which are held tightly, like calcium and magnesium, on the clay-humus magnet than the easily exchanged
cations. On the soil's clay humus magnet, the following
percentages of cations are desired: for potassium 2 to 5
percent, magnesium 10 to 15 percent, and calcium 65 to
75 percent. These are general figures. We will get
more exact for specific crops in the future. For the
most part, this sort of information still needs to be
obtained through research. /b4/

It is important to remember, Hylas, that no two plant
species, like no two animals, are alike. They have different requirements for growth. In order to optimize
growth, we need to know our plants well, like our best
friends. You see Hylas, just because your soil is fertile, doesn't mean your harvest will be large. You need
to fulfill all of a plant's growth requirements -
sunlight, proper temperature conditions, adequate humidity, clean air and not too much wind - matching the right
plant species to the proper climate.

One plant group which you should acquaint yourself
well with, Hylas, are those which fix nitrogen from the
air - primarily the legumes. This is the plant family
which has members like beans, peas, clever and alfalfa.
However, I really shouldn't say that legumes fix nitrogen from the air because they don't actually. Legumes
have a partnership, a symbiotic relationship with a
nitrogen-fixing bacteria, Rhizobium. Legumes and
Rhizobium bacteria mutually benefit each other. The
bacteria live in root nodules of legumes. The Rhizobium
take nitrogen from the air, fix it in their bodies and,
when they die, the nitrogen becomes available for the
legumes to use for their growth. In exchange for the
nitrogen, the legumes give sugars to the bacteria to
eat while they are alive in the nodules.

Compared to the number of plant species on earth,
there are few that have this nitrogen fixing symbiosis.
Most plants just can't get nitrogen from the air. They
have to wait until an adequate quantity of nitrogen

builds up in the soil for them to grow. Some plants require less nitrogen than others. Fortunately, due to the legumes' partnership with Rhizobium bacteria, they can obtain nitrogen from the air.

↭ Digestion ↭

"Ooooh woooh!" a voice sang out on the road. In a blink of an eye, Eatmore slipped back into the soil. He didn't even say goodbye to Hylas. "Ooooh woooh! Hylas!" the voice sounded again. "Yes! I'm over here!" shouted Hylas. "Oh, there you are. I've finally found you. Here's the kerchief you loaned me during my sneezing bout on the train. I forgot to give it back to you."

"Why Dahlia, I'm glad you found me. I was just speaking to a . . .," Hylas hesitated to say earthworm since Dahlia would think he was a little crazy, "a, a friend . . . in my sleep - you know, in a dream." "Yes, I can see from the matted down grass that you have been sleeping on the job. Can't take this hard farm work, eh Hylas?" she said jokingly. Hylas absorbed the tease without flinching. He changed the subject.

"Dahlia, how did you find me?" "Well Hylas, I went over to the general store and community center this morning. As I was querying about your whereabouts, a goat stepped in. He carried some tools home with you yesterday." "Gotaway," Hylas said. "Yes. He is a very congenial fellow. Gotaway told me how to get to your parcel of land. Here I am." Dahlia replied.

Their conversation paused for a moment; not knowing where to go. "How is your grandfather?" Hylas question-ed. "Oh, why he is just great - losing a little weight, but he gets around like a teenager. I think I will thoroughly enjoy living with him. The only problem is that his hearing is quite poor. I have to shout every-thing I say in his ear. Sometimes our communication becomes hilarious because he can't understand me.

"For example, yesterday he pulled some carrots out of the refrigerator. They were beautiful carrots from his garden. Some of them had started to spoil. Our conversation went like this: 'You shouldn't scrub your carrots before you store them,' I shouted. 'I did scrub them. Look at this spot. Some of them are going bad. I don't know what to do about them. There are so many carrots that I can't eat them all at once. I'm not a pig,' my grandfather replied. I shouted even louder, 'Store them dirty and wash them just before you eat them. If you get them wet, they spoil faster.' 'Yeah, I'll wash them better next time so they won't spoil so fast. Guess, I'll give some of them to our neighbors. These carrots shouldn't go to waste. They are such nice, big carrots,' he said admiringly. 'Yes, they are,' I agreed, and we left it at that.

"Really, my grandfather is a good man. I love him very much. Some say, he is tight with his money because he is so thrifty. I think of him as a conservationist – doesn't waste a thing. In fact, I think he is one of the wisest users of resources I know. He washes his clothes the old way because it saves water and soap, rather than use a machine. Also, he says, he can get them cleaner the old way. I think he is right. My grandfather isn't one who will spend large quantities of money on you, but he would do anything for you or anyone else. He is a very kind, old gentleman.

"One of his favorite pasttimes is to sit and tell you stories about his adventures. He always seems to come up with a new story; rarely tells the same story twice. On the occasion that he does happen to repeat a story, you know he is not pulling your leg because the tale has not changed over the years. His stories do not boast of what a grand fellow he was in his youth. Instead, they reveal some of the amazing predicaments and unusual people life has brought his way.

"You should stop by and visit him some day Hylas. I know you would enjoy speaking with my grandfather. I

believe you would not only be entertained, but would benefit from seeing him. He has farmed in this area all of his life. I'm sure he could give you some valuable advice."

The summer passed quickly. Hylas did learn many farming tricks and shortcuts from Dahlia's grandfather. Speaking to him was never easy because of his poor hearing, nevertheless grandfather always had an answer to any problem which Hylas posed. Sometimes, Hylas resorted to writing his questions down on paper. Hylas grew to welcome farming problems because he could use them as an excuse to visit grandfather, and in turn, see Dahlia. Even though the time he spent with Dahlia at her grandfather's house was as a threesome - grandfather never wished to be alone - Hylas still delighted in being in Dahlia's presence. The feeling was mutual.

As often as possible on sunny days, Dahlia would visit Hylas with a picnic lunch. Sometimes, she would help him in the fields after lunch, although Hylas rarely worked much when Dahlia was around. He was too enthralled with her. Fortunately, grandfather's advice helped to make up for the time that Hylas spent romancing when he should have been working. In fact, his crops were doing so well that the chickens who lived in the next parcel began seeking advice from Hylas.

When October arrived and the crops had all been harvested, it was time for the seasonal farmers to return for work in the city. Hylas and Dahlia were sad to part. They consoled themselves with the thought that they would spend a week together during the holidays.

Hylas was radiant as he rode the train back to St. Augustine. He was reflecting on how successful his growing season had been and how dearly he cared for Dahlia. Returning to the city would be a rather drastic transition for Hylas. Life in the country had been good for him. Hylas couldn't wait to begin a new growing season next spring. The question,"would he be happy in the city?" entered his mind.

"Of course he would," he thought. The city held
its own unique qualities and special adventures. He
anxiously looked forward to seeing his family and old
friends.

As Hylas stepped off the train at St. Augustine,
he was greeted by his brother, Rowan, and his six year
old niece, Picea. Rowan had a push cart which he had
brought along to carry Hylas' things in. Picea was
riding a new bike. She had received it for her birthday
- just two weeks ago. Hylas bounced over to Rowan and
gave him a big hug, picking him up off the ground. Then
he bent over and gave Picea a hug and a kiss. The little
girl never even got off her bicycle, acting as if she
would forget how to ride it if she got off. Picea glowed
with pride. "Well, happy birthday Picey. I see you got
a bike - a pretty red one. You even know how to ride it
already, huh? No training wheels even? Wow, how you've
grown over the summer."

"My dad taught me how to ride it, Uncle Hylas. The
hardest part was getting started. My dad showed me how
to use a step on the sidewalk to get my balance first,
then take off. I used to fall alot. See my knees."
They were all bandaged up. "Now, I can ride all the way
around the block without falling. You've grown too
Uncle Hylas," pointing to his beard. "Why yes," said
Hylas. "Everything grew prolifically where I was farming,
including my beard." "Well, let's be off," said Rowan.
"Haesel has dinner waiting for us."

The walk home was brief, only a mile and a half.
Along the way, Hylas noted how many more bicycles there
were now. The streets were covered with cyclists. "This
is rush hour traffic," Rowan explained. Many of the
bicycles were old models which had been refurbished.
The bike shops had obviously been busy.

According to Rowan, the bicycle shops had even offered bicycle maintenance classes for free. They reasoned to
their customers that a good bicycle will last almost
a lifetime if maintained well. This sort of class, you

might think, would tend to put the bike shops out of
business. However, the bicycle repairmen stayed busy
trueing wheels, making major repairs on bikes which had
been in accidents and refurbishing old bikes. The
classes actually encouraged business because they stimu-
lated people to buy parts, keeping their bicycles in
optimum working condition.

When the three rabbits walked through the door, they
were engulfed in the aroma of carrot pie and celery stew,
Hylas' favorite dishes. Haesel popped out of the kitchen
and lept at Hylas, giving him an extra big hug. "Why
Hylas, you don't look any different; except for your
beard. You don't even look like a farmer," she teased.
"Well, what do farmers look like Haesel?" "You know
Hylas. They have manure stained boots, old tattered
clothes which look as if they've been working in them
since the day they were born, hair that has been scorched
to a shade lighter and a ring around the crown of their
head from perpetually wearing the same, old hat." "Why,
I have a few of those qualities stuffed away in my
knapsack and hidden beneath my city clothes," he quipped.
"Oh, one more thing Hylas. That radiant look, as if
they have been soaking up the energy of the sun - like
a sponge absorbs water. You definitely have that
radiance so I guess you do qualify as a farmer."

"Tell us Hylas. How did you like farming? How
did you make out?" Rowan asked. "Oh, before you begin
answering questions, we better sit down to eat so our
dinner doesn't get cold," Haesel interjected.

"To tell you the truth, I love farming. Of course,
it is hard work, but my body thrives on exercise and being
outside. Also, I am extremely happy to be operating my
own place. I sort of feel like an artist because I
envision my painting, then apply it to my canvas, the
soil. As one would expect, the resulting creation is
never exactly what I envision, but that is what makes
farming stimulating. I am constantly being challenged
with new difficulties. Fortunately, I have made many

friends and collaborated with them to find solutions to
my problems.

"An old man, Dahlia's grandfather, has been particularly helpful in guiding me toward solutions. With the
aid of his advice, I have become one of the most successful seasonal farmers in our district. Now, other
seasonal farmers visit me for advice. I feel very good
about farming. What I plan to do with my land now is
improve the soil so I can grow crops which require high
fertility and lots of care - crops which other seasonal
farmers are not yet able to grow. This may sound boast-
ful, but I am making quick progress. I have natural
farming talent so I want to make the best of it."

"Well Hylas," started Rowan, "I hope you have as
much natural ability with carpentry, as you do with
farming because I want to offer you a job doing interior
finish work on apartments. They are restoring some of
the old buildings which used to be shops, turning them
into living spaces. It will be all indoor work. We'll
stay out of the cold. The outdoor work, like installing
solar collectors and putting in new windows was completed during the warmer months. What is left now is the
plumbing, electrical work and interior carpentry. I am
responsible for installing cabinets, doors, walls,
etcetera.

"You don't have to decide right now. I know there
are many alternatives for you to explore. Maybe you
already have something in mind, something like working
at a ski resort or at a factory building composting
toilets. There is plenty of work available for you returning, seasonal farmers. I thought it would be fun
for us brothers to work together. Like old times, when
we used to clean office buildings together. Geezuz, we
were efficient. Weren't we?"

"Yes, we were," replied Hylas. "Sounds good to me
Rowan, but let me look around a little before I give you
a definite answer. No sense in passing up any higher
paying job. I do appreciate the offer very much. Chances

are good that I will work with you. You know, I have
had some carpentry experience this summer, improving the
community center. I enjoyed it very much, so I would
probably be happy doing interiors with you."

"So Hylas, who is Dahlia?" questioned Haesel. "Is
she a new lover?"

"How perceptive of you Haesel. You've guessed
right. It must have been the way I said her name which
tipped you off. Yes, Dahlia is a beautiful woman whom
I've enjoyed being with this past summer. I miss her
already. You will get the chance to meet her during the
holidays. She plans to visit then."

Haesel started, "I'm glad to hear you are in love
again Hylas. I will look forward to meeting Dahlia and
hearing more about her as time goes on, but please
brother-in-law, no crooning around this house in the
meantime, while you wait for her arrival." Haesel loved
to jab Hylas with some teasing now and then. She couldn't
help it. It was as if she wished to remind him of the
ten years which separated them in age. "Don't worry
Haesel, I'll do my crooning only when I have pen and
paper in hand, not when I'm around energetic rabbits
like you." With this Hylas slowly rose from the table
and stated he was very tired from the train ride.

"Hey Picey, would you like to go to the arboretum
with me tomorrow - sort of a belated birthday present?
We could even ride our bikes." "Sure Uncle Hylas, I'd
love to go." Picey's eyes lit up. "Great! Then we'll
both get a good sleep and be off in the morning."

✌Indigestion ✌

The ride to the arboretum was a slow one. Fortunately, the arboretum was only one mile away from their house
and the road was not hilly. Picea rode very well for a
bunny her age. She had to stop and rest every few blocks,
but she didn't fall once. Hylas carried a pack with him
which had some food and additional warm clothing in it.
The streets were not as busy as they had been on the
previous evening. Instead of bicycles, the streets
were populated with leaves which danced and raced in the
wind.

It was a magnificent day to go to the arboretum. The
sky was blue and the air crisp. The trees even seemed
to enjoy the weather. Picea loved trees. Her Uncle
Hylas could not have chosen any better outing than a
visit to the arboretum.

They parked their bikes outside the gate. When
they entered the park, Picey took off running like a
racer. Hylas shouted with surprise, "Hey, not so fast
Picey. Wait for me." By the time he had finished saying
this, she was down the curved path and out of sight.
Hylas walked calmly along the path, trusting that his
young niece would stay out of mischief.

The trees retained only a small vestiage of their
summer clothing. All were quite bare, except for the
splotches of leaves left on the oaks. As Hylas turned a
bend in the path, he found Picey. "Look, Uncle Hylas!
My favorite tree, a Blue Spruce. Did you know that my
name is the latin name for spruce? We have the same
name. Isn't it big and beautiful?"

With this, Picey took off running again. "Wait Picey.

I want to walk with you." This time Hylas ran after her.
He caught up with Picea at her next stop, a Black Walnut.
"This is a Black Walnut, Uncle Hylas." Picey knew alot
about trees for a child of her age. She loved to profess
her knowledge to adults. They marvelled at her memory.
"You can eat the nuts of this tree. It looks like the
squirrels have already hidden them away for the winter.
I don't see any nuts around. Oh well, it is probably
better that we didn't find any. Last year, I ripped the
husks off of a pile of walnuts and my hands got stained
all brown. I couldn't get them clean for over a week."

"Will you run with me to my next best friend, Uncle
Hylas?" "O.K." he answered. "This is your birthday treat,
after all." Down the hill they bounced. "Look at the
bunches of leaves up in those trees, Picey. Hylas stopped running. "I wonder what they are? Could they be
eagles nests?" queried Hylas. "No, Uncle Hylas, we don't
have any eagles around here. Those are squirrel nests."
Hylas played ignorant. "Oh, so that's what they are.
What kind of trees are they in, Picey?" "Why, they are
oaks. They are all oaks. I don't see any squirrel nests
in the other trees, like the maples or sycamores." Hylas
looked thoughtful, "I wonder why they made their nests
only in oak trees?" Picey thought for a moment. "I know.
It is so they will be close to their food - acorns."
"Yes, that makes sense," agreed Hylas. They continued
running.

"Here we are," wheezed Picey, out of breath. "Do
you know what this tree is?" she asked. Hylas picked up
a leaf and examined it. "Oh yeah, I can tell by the
fat, four pointed leaves that it is a Tulip tree. It is
a beautiful tree, Picey. I enjoy its unusual flowers in
the spring. They remind me a little of Magnolia tree
flowers. Why Picey, here is my favorite tree, right next
to the Tulip tree. Do you know what that tree with the
long cigar shaped pods is?" Picey thought hard. "No, I
can't remember that one." "My favorite tree is a Catalpa," Hylas answered. "It has very large, heart shaped

leaves with gorgeous white flowers in the springtime.
The flowers have a wonderful fragrance. I believe that
is why I am so enthralled with this tree."

   Because they were looking up, Hylas and Picey hadn't
realized that a well bundled up rabbit in a dark, blue
coat was sitting against the Catalpa tree. "What a
coincidence, this is my favorite tree too," a woman's
voice said.

   The woman turned her head to face Hylas and Picey.
"Why Violet, is it really you?" Hylas said with surprise.
He went over to the Catalpa tree and gracefully helped
Violet stand up by pulling her paws. Violet gave Hylas
a big, long hug. She was an old love of Hylas'. They
had ended their romance when she decided to go overseas.

   "Yes. It's me, Hylas. You remember me, after all
these years, eh? I certainly haven't forgotten you. "Why
Violet, the last I heard, you were still in the tropics.
When did you get back?" "In the middle of August," she
replied. As you probably know, I got stuck in the tropics
during the Fall. I was unable to return to St. Augustine until this summer. My love for the tropical jungle
has not diminished, but I became anxious to return
home, for a visit at least. You know how a person
usually has the strongest desire for what is out of
reach, Hylas." "Well, I'm glad you're back Violet."

   "Oh Violet, I'd like you to meet my niece, Picea.
Picea is very fond of trees. In fact, she knows more
about them than I do. Her birthday was last week so we
are on a birthday adventure today." "How nice Hylas –
I'm happy to meet you Picea," said Violet. "Picea,
this is a very good friend of mine, Violet." Picey
shook her hand with little interest, wishing to continue
running with her Uncle. "Would you like to walk with us,
Violet?" "Yes, I would enjoy that Hylas. We can catch up
on each others activities over the past few years."
Exactly what Picey feared was now happening. She had to
share her Uncle Hylas' attention.

   Much to Picey's surprise, she actually had fun with

Violet. Violet knew more about trees than her Uncle.
Picey was hungry for more tree lore. There were few
questions which Violet could not answer. Picey was
also delighted with Violet's stories about tropical
plants and creatures. As for Hylas, he was thrilled to
be with his old lover again. They had had an exciting
and adventurous relationship. It was as if they had
been two pieces: of a puzzle. Their lives were much
fuller when they were together. It was sad when they
parted, but the desire to part had been mutual. Both
needed more adventure. Adventure they could obtain
more easily on their own.

Violet invited her two hiking companions over for
lunch. Hylas and Picea gladly accepted. After lunch,
Picea fell asleep. This gave Hylas and Violet the opportunity to investigate their refound relationship.
They enjoyed each other's company as much now, as they
did when they were younger. When Picey woke up, the
sun was starting to set. Hylas and Picey bundled up
and set off for home on their bicycles. it had been
a remarkably memorable day for all three rabbits.

Hylas and Violet continued to spend time together
in the following months. As the holidays approached,
Hylas' thoughts turned toward Dahlia. He had written to
her regularly since leaving the farm. Now, Hylas was
afraid that things might get sticky during Dahlia's
visit. "A friend, a woman friend, is going to visit for
a week during the holidays. I will be tied up most
of the time." he told Violet. Violet took the news well
- almost too well.

Dahlia arrived at the St. Augustine train station
on the day after Christmas. She was full of love and
longing for Hylas. "Oh my dear Hylas, how I've missed
you these past two long months. Never before have I
counted the days till Christmas with such expectation.
You were in my thoughts every day. All my meals have been
memorable dreams of you. Every mouthful of carrot or
potato from your farm kindled a flame of longing to be

with you. The day I have dreamed about for so long is finally here."

Hylas was equally enamoured with Dahlia, yet, as he hugged her, he could not help but think: "What am I going to do about my two loves? What if they meet? Who do I love more?" Hylas quickly pocketed these thoughts, after rationalizing that he needn't make any decisions now. He should just enjoy himself. After all, he wasn't married. "My love and affection is equally strong for both Dahlia and Violet." He left it at that.

During the two months they were apart, Dahlia had crocheted a beautiful, wool-mohair scarf for him. You could easily tell by the look on Dahlia's face, as he opened the package, that she was with him all the while she worked. Every loop was another kiss on his forehead. The completed garment would be a lasting memory of her arms around him.

All this love was beginning to make Hylas feel guilty. Dahlia's love was so syrupy sweet that Hylas felt the onset of obesity. He didn't know what to do. Hylas felt like he was hiding Violet from Dahlia. But, was he really? Why should he tell her about Violet? Dahlia would only be in town a short time. Why should he complicate their relationship which was so unimaginably good?

Dahlia and Hylas had a wonderful time exploring the sights of the city together. Hylas' work as a carpenter with his brother had kept him quite busy till now. He never had the opportunity to fully immerse himself in the entertainment which the city had to offer. Fortunately, he was able to take a week off of work to enjoy with Dahlia. They went to a concert, a play, museums, art galleries and restaurants with delicacies from almost every part of the world. Hylas was flaunting his hard earned money. He earned it. Dahlia loved the pleasure of expensive leisure, but she would have been satisfied spending time with Hylas anywhere.

Two days before Dahlia's departure, Hylas could

stand it no longer. He had to tell Dahlia about Violet.
There must be complete honesty between them, or he would
have no peace. They had just returned from a romantic
dinner at a French restaurant. "Dahlia, you know that
I love and care about you more than I can say. You are
the light of my life. You know I never want to hurt you.
Because of this love, I want to be completely honest with
you. I have been seeing an old girlfriend for the past
two months. We ran into each other accidentally and
I . . . well, just fell into spending time with her again.
This hasn't changed my feelings toward you one iota."

Dahlia's mouth dropped with surprise. Her smile
disappeared. "That's okay, Hylas, I understand."
Dahlia's eyes were becoming watery. She was on the verge
of bunny tears. Hylas grabbed Dahlia and comfortingly
held her in his arms. His heart sank so low that he felt
it would burst its way out of his big toe at any second.
Had Hylas made a mistake telling Dahlia? He now thought,
he had made a big one.

"Dahlia don't feel sad. This really doesn't change
our relationship at all. She's just an old friend. In
fact, Violet, that's her name, will probably be leaving
for the tropics before you know it. She is a world
traveler. Violet loves adventure far more than she loves
me. We just happen to spend time together lately, that's
all. Think of her as my friend, as if she were a male."
Hylas tried hard to lighten the blow of this news.
"Alright Hylas, I said I understand. You live your own
life. I'm not your ball and chain. I know this hasn't
changed our love at all." Yet, deep inside they both
knew it had.

The next morning, Hylas decided they had spent
enough time inside buildings. It would be far more
enjoyable to be outside with Dahlia, even though it was
cold. They were both hardy, young rabbits who loved
nature and sport, so they decided to go skiing. Dahlia
was rather slow and somber along the trails. She had lost
her zest. Hylas teased her on by challenging her to a

race. After much coaxing, Dahlia complied. As they
were zooming like daredevils down a rather steep slope,
another snow bunny in a bright green jacket skied out in
front of them. Dahlia couldn't stop fast enough. Down
the two rabbits tumbled.

"Oh no!" Hylas shouted as he looked back. Quickly,
he turned around and skied over to them. "Are you all
right?" Hylas said meaning both of them. "I'm miracu-
lously okay," sneered the green skier. "I am too," said
Dahlia. "That's the last time I race with you, Hylas. I
knew I shouldn't have let you talk me into it." She was
venting some of her anger from the previous night at him.

Hylas ignored Dahlia's tone. He was now faced with
a bigger problem. As Dahlia was speaking, the green
skier had removed her goggles, hat and coat in order to
clean the snow out of the back of her neck. The green
skier was Violet.

"Well . . . " Hylas hesitated before saying her
name, "Violet, it's you." At this point, Hylas figured
that his luck was on vacation so he would just let what
happened happen. "What a funny way to meet," Hylas joked.
"Dahlia, this is Violet. Violet this is Dahlia." "I'm
pleased to meet you Dahlia," said Violet. "Even if it is
under such unusual circumstances. Hylas has told me
many good things about you." Hylas' ears picked up. He
hadn't told Violet anything about Dahlia. "Yes, I've
heard about you too, Violet. You are the world traveler,
who loves the tropics." Much to Hylas' surprise, the
two ladies hit it off quite well, almost ignoring Hylas.
The three rabbits agreed to ski back to the lodge and
have a hot drink together.

That evening, Hylas rode home in a sleigh with
Dahlia. She had made a complete change in attitude
from that morning. Dahlia was affectionate with Hylas
once again. "You really liked my friend, Violet, huh
Dahlia? I thought you would." He acted confident, as if
he had had everything under control all day. "This
morning, I thought you were mad at me." "I told you I

understood, Hylas. I do even better now, after meeting
Violet . . . " pause. "You silly toad - you shouldn't
worry so. Violet is much too wild for such a deadbeat as
you." Hylas said nothing. He was stranely relieved.

* * * * * * * *

"Now, isn't that a horrible thing to say to anyone?"
Hylas asked Eatmore. "Actually, I think she was just
teasing you to lighten the situation. Are you still
seeing these two women?" "Why yes," said the rabbit.
"Well, I wouldn't worry about it then. Just live.
Live truthfully. All will go well."

Hylas ran into, or should I say, almost ran into
Eatmore, just as he was turning the last few forkfulls
of his compost pile. It was spring again. All the
successful seasonal farmers had returned to the country.

"Hylas, I'm upset with you. You have a bigger
problem with me than with those flossies. I thought you
agreed to feed the soil properly when I gave you that
exhilarating lecture, or should I call it pep talk, last
spring. Hylas, you didn't even plant a winter cover crop
before you left last fall. Haven't you heard of green
manure? What do you expect us soil creatures to eat and
live in this season? Did you think you would haul
manure for this whole five acres? Where would you get
it? I just don't understand you, Hylas. After that
long lesson I gave you, I thought you knew what you were
doing. Maybe, it's me? Maybe, I wasn't clear enough?
Maybe, I'm not as good a teacher as I thought I was?
Well Hylas, what do you have to say for yourself?"

Hylas had gotten jittery after this reprimand. Up
till now, all he had received was praise for his farming
ability from everyone else. He stood on one foot, then
the other, acting as if he would love to hop away from
the frightening words which were barking from Eatmore's
little mouth.

"Eatmore please, I beg your pardon. Really, I think
you are a very good teacher. I understood everything you
told me last spring, but, honestly, I do not remember
your mentioning anything about green manure or cover

crops. Why, I don't even know what they are. Green manure sounds like the punch line of a joke to me."

"Enough, enough. Okay Hylas, you didn't know. I guess, it is my fault. In my haste to dig deep tunnels for overwintering last fall, I must have forgotten to tell you about green manure. Well, sit down and relax my friend. I'll tell you all I know about green manure. Once you know about green manure, you will never forget again."

Green Manure, a Natural Plant Food

"Everyone knows that manure is an excellent fertilizer. Well, green manuring uses green plants for the
same purpose - a crop still green and growing is plowed
under to improve the soil. It is a natural soil and
plant food. Cover cropping is when you sow seed to protect the soil from erosion." "Oh yeah, I've done that before," said Hylas, "in landscaping". "Green manuring is
the last step in cover cropping. You may not have heard
of green manure, but you are using it whenever you plant
a cover crop and turn it under. The name may be new, but
not the practice.

Cover cropping and green manuring follow nature's
way of holding the soil and maintaining its fertility.
A green manure crop provides a ground cover which holds
the soil and prevents erosion while it is growing. Once
the green manure crop is turned under, it adds organic
matter to the soil and insures nutrients for succeeding
crops. The best plants for green manuring are the
legumes because, along with organic matter, they contribute vital nitrogen to the soil.

Organic matter is literally the life of the soil.
The annual decomposition of organic matter, through the
activity of soil organisms, makes nutrients available for
vigorously growing plants. Organic matter also lightens
the soil, to promote better aeration, drainage, and
moisture retention. But, the loss of organic matter is
a constant threat to cultivated soils. Many factors
lower the fertility and productivity of cultivated soil
by exhausting organic matter. You expose organic matter
to wind, air and sun when you work the soil, the growing

crop uses it, you remove the crop's potential return of organic matter to the soil when you harvest, rain and irrigation water leach away nutrients supplied by decomposed organic matter, and, finally, erosion washes away organic matter and the soil itself.

Soil is a precious resource. The natural process of soil formation is a very slow one - it has taken ages for the bare primordial rocks, through weathering and the gradual accumulation or organic matter, to become the loam in your field. Nature carefully protects this precious resource with a blanket of vegetation. We can disregard these fundamental principles of soil formation and retention only at the cost of long-term degradation of the land. When we work against nature, fertility falls, erosion accelerates and, finally, agriculture becomes nonproductive. To sustain productivity only by temporarily supplying plant foods, such as chemical fertilizers, is to lose sight of the life of the soil. A viable agriculture - in which the land is productive year after year, decade after decade - must look to the soil itself, to conserve and maintain itself.

There are many ways to maintain and improve the soil by adding organic matter to it. Green manuring is not the only method. Leaving the land 'fallow' or unseeded - allowing the weeds to grow - for a year or more is how farmers originally restored fertility to the land. Plowing weeds under adds a good deal of organic matter to the soil. You don't have to let your land lie fallow for a full year to take advantage of weeds. A season's growth of weeds can be a plus rather than a minus if you can turn them into green manure. However, certain weeds may take up more than their share of ground moisture and nutrients when growing with a crop. In order to minimize future weed problems, you should plow weeds under before they go to seed.

Crop rotation replenishes the soil without skipping an annual crop as in fallowing - an obvious economic advantage. For example, nitrogen-consuming corn and

nitrogen-fixing clover are often rotated. After the
nitrogen-fixing legume has been harvested or grazed,
the stubble can be plowed under. The stubble, along
with the organic matter and nitrogen nodules in the
root system, help to improve the soil's fertility.

Both fallowing and rotation are beneficial to the
soil, but green manuring is a faster and more effective
way to maintain and enhance soil fertility. It is a
faster and more effective method because you can choose
specifically the green manure crops which will produce
the most organic matter and nitrogen from your special
soil and climatic conditions. Most of the time, you will
choose a green manure crop that will not interfere with
your cash crop. A green manure can grow before, after,
or around the cash crop so you do not lose income by
skipping the cash crop.

Green manuring is an ancient farming technique. The
Chinese practiced it perhaps three thousand years ago,
as did the Greeks and Romans before the time of Christ.
Even though it has been used for millennia, green manuring
is much less common in today's agriculture – due to the
introduction of chemical fertilizers. It _is_ easier to
apply a chemical fertilizer than to grow a green manure
crop and turn it under, but chemical fertilizers deplete
the soil in the long run. In addition, the resources
needed to manufacture chemical fertilizers (e.g. natural
gas and mineable phosphorus) are becoming scarcer and more
costly. Green manuring is destined to gain more common
usage.

⌇ Green Manure
in your Rotation ⌇

There are four ways to use green manure:
1) Main Crop: a green manure crop that occupies the land for the entire growing season.
2) Companion Crop: a green manure crop intermixed with the cash crop, but susidiary to it.
3) Catch Crop: a green manure with a short growing period to fill out the growing season <u>before</u> a late cash crop or <u>after</u> an early one.
4) Winter Cover Crop: a green manure crop that occupies the land between growing seasons.

The manner in which you will use each of these four methods of green manuring depends on whether your cash crop is an annual or a perennial plant. Annual plants are those which grow, flower, produce seed and die, all in one year. They reproduce by seed <u>only</u> and are never woody. Each year, an entirely new plant sprouts up from seed. Perennial plants can continue to grow, flower and produce seed year after year. They can sometimes reproduce, not only from seed, but also from their roots (rhizomes) or other plant parts, like runners (stolons) as strawberries do. Many perennial plants are woody. Some of the temperate perennials drop their leaves in the fall.

There is a third kind of plant, a biennial. Folks often forget that certain plants are biennial and treat them as annuals. You will find it valuable to know what a biennial is and whether a plant is a biennial. This way, you can avoid unwanted surprises. Biennial plants grow for two years, flowering and producing seed in the second year. Normally, they only reproduce by seed.

Sometimes, biennials can act like annuals, growing only
a short time; or perennials, growing longer than two
years. This depends largely on whether the growing
conditions for the plant are favorable or not.

Annuals	Perennials	Biennials
tomatoes	apples	carrots
cucumbers	oranges	sweet clover
corn	rhubarb	red clover
cotton	alfalfa	
soybeans	white clover	
vetch	Sericea lespedeza	
field peas		

1) Main Crop:
   Annual Cash Crops — When the soil is so poor that
you can't grow any cash crop, green manuring for an
entire season can increase the fertility of the soil
enough to establish a regular rotation of cash crops.
For example, in North Central Europe, spurry has been
grown to rejuvenate unproductive, sandy soil. Spurry
will grow in dry, sandy soil where clover will not grow.
This green manure crop grows so rapidly that you can
grow and turn under three successive crops in one season.
Once you have grown and turned under spurry for one
season, the renovated soil is now fertile enough to
produce a winter grain crop or clover. Similarly,
you can turn under several crops of buckwheat in one
season to rejuvenate the soil. /a5, w3/
   Yellow lupine is another green manure which is
valuable in restoring sandy land. However, unlike
spurry and buckwheat, it is a legume. Yellow lupine
adds nitrogen to the soil, but you can only turn under
one crop per season. A main crop of yellow lupines can
restore soil fertility so a rotation with potatoes and
winter cereal, for example, can be established. /d1/
   Any green manure which will grow on poor, infertile
soil can be used as a main crop to restore fertility.

Usually, it is preferable to grow a legume, like Crotolaria, Dalea, Sesbania, velvet beans or cow peas, as your main crop to add nitrogen along with organic matter. However, in some cases, you may wish to increase the soil's organic matter content quickly, before sowing a legume.

Another way to use a main crop green manure is to grow a perennial hay or pasture crop for a few years. Then, during the last season, allow the vegetation to grow up and turn it under as a green manure. A farmer in Kansas did this effectively with alfalfa. His soil was worn out. The alfalfa grew poorly the first year. The following two years, the alfalfa took off and he harvested several cuttings of hay. In the third year of growth, he allowed the alfalfa to regrow just before fall and plowed it under.

Alfalfa pumps up nutrients from the subsoil. In this situation, the tilled under alfalfa provided nutrients for a corn crop the following year. Sainfoin is another green manure which can be used in this manner. Its roots penetrate 2 to 3 times deeper than alfalfa. Like alfalfa, sainfoin is a good hay plant. /r4/

Perennial Cash Crops - A main crop green manure can also insure a good start for permanent plantings - in establishing an orchard or vineyard, for example. Let's say, it is late in the spring and you just bought some land. You want to grow grapes on this land, but it is too late to plant them. The soil is quite poor anyway, so you decide to: first, have your soil analyzed, second, add soil ammendments and third, plant a green manure crop.

You decide to sow sweet clover because it is a legume with a deep root system that is well adapted to your soil and climate. Sweet clover as a green manure will add organic matter and nitrogen to the soil. Its root system will penetrate deeply into the soil, making passages for the forthcoming grape roots. Additionally, the sweet clover roots are able to extract phosphorus

and potassium from insoluble minerals in the subsoil,
thus bringing these nutrients up to the topsoil for use
by the young grapes. The following spring, you turn
under the sweet clover just before transplanting the
grapes. As it decomposes, it feeds the developing grape
vines. The grapes get off to a good start. Now you can
start thinking about a Companion green manure crop for
the grapes.

2) Companion Crop:

   Annual Cash Crops — A leguminous green manure can
sometimes be intermixed with a cash crop as a companion.
Quite often, there is a good deal of space between the
rows of a cash crop. A companion green manure can be
grown between the rows; then be turned under after the
cash crop is harvested. The advantage of doing this is
that the green manure can donate nitrogen to the row crop
while they are both growing. You do not have to give up
a growing season for a green manure. Instead, you can
still harvest a cash crop. If you are mixing a companion
crop in with a cash crop, make sure that the cash crop
gets enough water and space. Remember, the companion
crop is subsidiary to the cash crop. With care, you can
assure that both crops have enough water and space.

   Some examples of this kind of intercropping are:
peanuts grown with corn or millet; and mung beans with
sorghum or cotton. In some situations, you can sow the
companion crop after the last time you cultivate the
cash crop; such as sowing crimson clover in corn,
then turning both the clover and the stubble under after
you harvest. /m4, s3, s7/

   Perennial Cash Crops — In perennial cash crops,
such as in orchards or vineyards, it is beneficial to
plant a companion green manure under the trees or vines.
Some of the perennial grasses and legumes are suitable
for this purpose. You can grow a perennial companion
crop under apples, for example, and mow this green
understory. The decomposing clippings feed the soil.
You could grow annual legumes and turn them under if

your trees are in poor soil and need extra feeding. Two
examples of companion crop mixtures for use under a
perennial cash crop are: bluegrass, white clover and
crown vetch; or strawberry clover and orchardgrass. All
of these species are perennial.

3) Catch Crop:

Green manure as a catch crop fills out whatever
part of the growing season a cash crop does not occupy.

Before a Cash Crop - Before a late planted cash
crop, such as potatoes or soybeans, you can plant an
early spring catch crop, like oats and field peas, for
green manure. Turn under the green manure three weeks
before planting the cash crop.

After a Cash Crop - More often, a catch crop follows
an early harvested cash crop, such as winter wheat or
oats. (Notice that oats can be grown for either a green
manure or a cash crop.) In mid or late summer, you can
plant lupines, mustard, rape, turnips, Tepary beans, cow
peas, soybeans or buckwheat for green manure. If only
a very little of the growing season is left, you can
plant sudangrass, or pearl millet in the South.

Sometimes, you can start the catch crop before the
cash crop is harvested. Crotolaria, for example, can be
sown into oats in the late spring. Likewise, Annual
Lespedeza can be sown with oats and other early harvested grains, but they should be sown in late winter or
early spring. Both of these legumes make slow growth in
the spring and rapid, sizeable growth in the summer.

In planting a catch crop, you want a crop that will
make the most growth in the time available - whether early
spring or late summer. If possible, you should choose
a legume before a non-legume because it will add nitrogen
along with organic matter to the soil.

4) Winter Cover Crop:

Winter cover crops are perhaps the most important
use of green manuring. They are important because, on
most farms, a cash crop is not grown during the fall and
winter. The winter is an ideal time to grow a soil

SOME EXAMPLES OF GREEN MANURING SYSTEMS

GREEN MANURE TYPE	1st Year	2nd Year
Main Crop	3rd Year Alfalfa / Sweet Clover	$ Corn $ / $ Grapes $
Companion Crop	$ Cotton $ / Mung Beans / $ Corn $ / Crimson Clover	
Catch Crop	$ Apples $ / Strawberry Clover & Orchard grass / Oats & Field Peas / $ Potatoes or Soybeans $	$ Winter Wheat $ / Yellow Lupines
Winter Cover Crop	Bur Clover / Rye & Hairy Vetch	$ Cotton $ / $ Tomatoes $

improving green manure crop. If a winter cover crop is
properly timed, your summer cash crop will not be interfered with. The crop you select as a winter cover and
green manure depends upon the severity of the winter
in your area.

Annual Cash Crops - In the South, bur clover has
proven to be a valuable winter cover crop for cotton
fields. Growing bur clover in the winter is an inexpensive way to boost soil fertility for cotton; especially
since bur clover will reseed itself when allowed to
mature seed. In the North, winter hardy green manure
species, like hairy vetch and rye, make a good mixture
for a winter cover crop. Such a mixture can be grown
between crops of tomatoes, for example, improving the
soil as they decompose in the spring and summer.

Perennial Cash Crops - In dryland orchards - orchards
in regions with dry summers and little or no irrigation -
the growers keep the soil cultivated clean to prevent
any weeds or understory vegetation from competing with
the trees for water. They feed the soil by growing a
winter cover crop which is turned under in the spring.
A good example of such a system is in citrus groves.
Sour clover produces very well as a winter cover crop in
citrus. It is invaluable for feeding the citrus trees.

The Benefits
of Green Manuring

"I've touched on some of the benefits of green manuring, Hylas. Even so, I want to illustrate the benefits further. This way, you will understand me more clearly and this information will stick in your head.

Green manuring increases soil fertility by adding organic matter and humus, promoting soil life, adding nitrogen, pumping nutrients up from the subsoil, improving soil structure and drainage, preventing soil erosion and leaching of nutrients, and, finally, detoxifying the soil. Insect and weed problems are also reduced through green manuring. With all of these advantages, you can understand why I got upset with you. Green manuring is too important a practice to forget.

Adds Organic Matter and Humus

The most valuable benefit of green manuring is that it adds organic matter to the soil. Organic matter is the life of the soil because it is the basic foodstuff of soil organisms. The soil organisms feed on organic matter and change it into a form which plants can use. This changing, or transformation of organic matter is what makes soil fertile, not just the mere presence of organic matter. The soil must have organic matter, along with soil organisms, in order to transform organic matter into plant food. Soil must be alive to be fertile. /p5/

When organic matter is eaten and broken down by soil organisms, they transform it into humus, carbon dioxide, organic acids and other molecules. The molecules which we are most interested in are the simpler molecules which contain the plant nutrients often in short supply, like nitrogen and sulfur. You will remember, nitrogen and

sulfur are mobile nutrients. They are highly water soluble and travel readily in the soil. As organic matter is
gradually decomposed by soil organisms, molecules which
contain nitrogen and sulfur are released - a timed release. These two important nutrients become available
to plants in small quantities as organic matter breaks
down. The plants continuously absorb nitrogen and sulfur
from the soil solution and use them to build their tissue.
As the content of organic matter increases, the reserve
content of nitrogen and sulfur increases since they are
important constituents of organic matter.

Now, the carbon dioxide and organic acids which
result from the breakdown of organic matter are also
important in providing plant nutrients. The carbon
dioxide mixes with water to become carbonic acid -
the bubbly stuff in soda pop. Both, carbonic acid
and organic acids flow down into the soil and dissolve
nutrients from the rocks, making these nutrients available for the plants to use. One particularly valuable
nutrient which is liberated in this way is phosphorus.
/f3, g5, p5, r3, t1, w4/

The most valuable product resulting from the decomposition of organic matter is humus. Humus is derived
from the cell structure material - the plant parts more
resistant to decomposition - and carbohydrates. Generally, humus can be considered to be the conglomeration of
molecules which cannot be broken down any further by soil
organisms and they are not water soluble. This conglomeration of leftover molecules, from the meals of the soil
organisms, make up very large humus molecules. Humus
molecules are chemically active. They, like clay, have
a large number of negative charges on their surfaces.
These negative charges attract and hold positively
charged plant nutrients, cations such as: potassium,
magnesium and calcium. They hold these nutrients until
the plant roots come in contact with them and absorb
them.

Humus also reacts with clay. They get linked up to

form a clay-humus magnet. Sand and fragments of silt,
on the other hand, are not chemically active. They aren't
negatively charged and do not hold cations for plant roots
to absorb. Consequently, organic matter and humus are
even more important in soils high in sand or silt.

Some clays are stronger magnets than others. That
is to say, they can hold a larger quantity of cations.
For example, montmorillinite (expanding clay with a high
CEC) holds more cations than kaolinite (non-swelling clay
with a low CEC). A sizeable humus content would be
even more important for soils with kaolinite than montmorillinite. /r3, w6/

When the clay-humus magnet is not present in a soil,
the positively charged nutrients, like potassium, magnesium, calcium and ammonium, get washed down and away by
rain or irrigation water. The cation holding magnet is
essential to soil fertility. If the clay-humus magnet
is not operating, then positively charged nutrients get
washed away. They go into the water system rather than
becoming available for plant growth. Chemical NPK fertilizer won't even work properly without the clay-humus
magnet. They give just a quick shot of nutrients and
then flow away. But, when plenty of organic matter is
supplied to the soil and the clay-humus magnet is strong,
chemical fertilizers are no longer necessary.

Listen to me Hylas. You must add organic matter to
the soil annually to feed the soil organisms which in
turn feed the plants and to keep the clay-humus magnet
strong. Frequent additions are better than one large
feeding. Remember how nature does it - the leaves drop
from the trees every fall. Likewise, the plants flop
over. Both feed the soil.

Green manuring will become an even more important
technique tomorrow, Hylas, because organic matter will not
be as easy to obtain in the future as it is today. As the
earth's population increases, more crops will be eaten
directly, rather than be fed to animals. This will mean
a decrease in available animal manure. You can easily

see how important green manuring will be as a source of
organic matter in the future.

Or, what if you gardened in the city, instead of
farm out here? Manure isn't readily available in the
city. You will need to use green manure to promote and
sustain the fertility of your soil. Green manuring is
better than adding animal manure to the soil anyway.
First, if you feed legumes to animals, rather than directly to the soil, only 2/3 of the nitrogen fixed by the
legumes is gained by the soil through the animal manure
- a sizeable loss. Second, green manure is a more econ-
omical source of organic matter. Unlike animal manure,
there is no time and work in collecting and hauling.
Instead, you can grow your organic matter in large amounts
right where it is needed. /a5/

Promotes Soil Life

Not only does green manure act as a source of food
for soil organisms, but it also improves the soil habitat
for them. Green manure acts as an insulating blanket
which keeps the soil warmer in the winter and cooler in
the summer. The soil doesn't go through such a wide
range of temperature fluctuations.

A blanket of green manure particularly encourages
earthworm activity. Earthworms, like myself, are especially beneficial to soil fertility and plant growth
because we burrow deep into the subsoil. The channels
we dig become passageways for plant roots. We also bring
nutrients up from the subsoil to the surface, where the
plants can use them. /a5, f3/

Another type of worm, called an eelworm, or nematode,
is not your friend. Nematodes are not your buddies
because they feed on plant roots. Green manuring helps
to eliminate this pest by providing a suitable environment
for nematode trapping fungi. These predaceous fungi are
amazing creatures. One type has cells which form a
noose. When a nematode slithers through the noose,
the fungus catches and eats it. Organic matter provided
by green manuring promotes the growth and reproduction

of beneficial organisms like the nematode trapping fungi.
/a8, y1/
     Other good guys which green manuring will promote are
the free living, nitrogen-fixing bacteria. These bacteria
add nitrogen to the soil, but they do not have to live in
the roots of legumes. Like so many other soil organisms,
the free living, nitrogen-fixing bacteria feed on organic
matter. Green manure is an especially good source of
food for them and promotes their growth. Azotobacter is
one type of free living, nitrogen-fixing bacteria which
has demonstrated a great deal of promise for adding
nitrogen to the soil. I'll tell you more about this
microorganism later.

Adds Nitrogen

     When a legume, like clover, peas, vetch or alfalfa,
are grown for green manure, the soil is benefitted not
only by adding organic matter, but also by adding nitrogen. Nitrogen is fixed from the air into legume root
nodules by the nitrogen-fixing bacteria, Rhizobium.
     The nitrogen fixed by Rhizobium may be used by
plants in three ways. First, it may be used by the legume host plant. The legume benefits greatly by the
symbiosis, enabling it to grow in nitrogen deficient soils
where many other plants cannot grow. Second, the nitrogen
may pass into the soil itself, either by excretion, or,
more probably, by the sloughing off of the roots and
nodules. A crop grown with a legume may thereby benefit.
For example, a bluegrass pasture grows vigorously when
grown in association with clover. Third, when a legume
is turned under as green manure, nitrogen becomes available to the succeeding crop, improving its growth. /b3/
     You will be surprised Hylas, at the quantity of
nitrogen which some legumes add to the soil. Here are
some examples:

	Lb.s/acre of Nitrogen
Alfalfa	194 - 251
Alsike clover	119 - 141

Lb.s/acre of Nitrogen	
Annual lespedeza	85
Crimson clover	94
Cow peas	90
Field beans	40 – 58
Hairy vetch	68
Kudzu	107
Ladino clover	179
Peanuts	42
Red clover	114 – 151
Soybeans	100 – 105
Sweet clover	119 – 168
Vetch	80
White clover	103
/b3, t1, w7/	

I'm sure you'll be even more enthusiastic about legumes as a source of nitrogen Hylas, when you find out that it takes the energy equivalent of 60 gallons of gasoline to produce, transport and apply 100 lb.s of chemical nitrogen fertilizer. /s10/

Chemical nitrogen fertilizers will soon become a near-extinct technology because they are produced with fossil fuel. It is not a sustainable technology. There is a finite quantity of fossil fuel on the earth. Someday, the deposits of fossil fuel will be all used up. Most of us have forgotten that fossil fuel agriculture is very young – much less than 100 years old. This is barely the blink of an eye in agricultural history.

Legumes, on the other hand, provide a renewable source of nitrogen. They can be planted year after year, indefinitely. Agriculture can be sustainable through good management, ingenuity and care. Legumes make this possible.

Natural gas is the fossil fuel used to manufacture chemical nitrogen fertilizers. The price of natural gas will continue to rise because it is a finite resource. Its supply will gradually dwindle as time goes on. As

the price of natural gas increases, so will the price of
chemical nitrogen fertilizer. Due to the rising cost of
chemical nitrogen fertilizer, green manuring will appear
more and more favorable in the future.

The worst aspect of chemical nitrogen fertilizers
is that they do not add organic matter to the soil. They
do not feed the soil. In fact, they often kill soil organisms, like earthworms, and inhibit the nitrogen fixing
capability of legumes. We soil organisms are your unseen
workers under the soil surface. You can help us do a
better job of feeding the plants by feeding the soil.

Now Hylas, when you are deciding whether to harvest
a legume or turn under the whole plant, you must consider
where the nitrogen is in the legume. If a legume is
allowed to produce seed, most of the nitrogen assimilated
during growth passes into the seeds. When a legume is not
allowed to produce seed, most of the nitrogen is in the
vegetation of the plant. Here is a comparison of five
legumes, showing the percentage of nitrogen in their tops
and roots:

	Tops	Roots
Soybeans	93%	7%
Vetch	89%	11%
Cow peas	84%	16%
Red clover	68%	32%
Alfalfa	58%	42%

/k1/

You can see Hylas, less nitrogen is added to the soil
when the seed or hay is harvested and the roots or stubble
(also known as aftermath) is turned under. However, for
legumes, like red clover and alfalfa, you would still
add a considerable quantity of nitrogen to the soil if
you just turned under the roots. For any harvested legume, the roots still contribute some nitrogen to the
soil. Any time you include a legume in your rotation,
you benefit the soil with additional nitrogen.

When you can turn under the vegetation of a legume,

your succeeding crop will respond even better from the
additional nitrogen. For example, wheat grown where
soybeans were cut for hay give a much smaller yield than
when the soybeans are turned under. However, most of us
don't want our leguminous green manure to interfere with
the growth of our regular cash crops so we grow them as
companion crops, catch crops, or as winter cover crops.
Vetch and Field peas are two of the more popular winter
cover crops. They can fix free fertilizer for you all
winter, while your fields would normally be empty. /h4/

Pump Nutrients Up From the Subsoil

Many green manure crops, such as alfalfa and sweet
clover, have deep root systems. Their roots absorb subsoil nutrients and transport them up into their foliage. When the foliage of these deep rooted green
manures decompose, the subsoil nutrients become available
for shallow rooted crops. Deep rooted green manures
help provide nutrients for shallow rooted plants.

The roots of some green manures can even extract
nutrients from insoluble rocks. Rye, buckwheat and
sweet clover, for example, can extract phosphorus
from some insoluble minerals. These green manure plants
can be extremely valuable in boosting phosphorus availability in soils which have plenty of phosphorus, but
in an insoluble mineral form. /a8/

Improves Soil Structure and Drainage

Clay particles, microscopic in size, come in many
shapes and sizes. Some are shaped like plates, some like
blocks, and others like columns or granules. Any specific type of clay has symetrical and fairly uniform particles. They can fit together tightly, like bricks; especially when the clay particles are compressed together
by a heavy object, such as a tractor, or when the clay
dries out. If clay particles become packed together
tightly, then there isn't any space for air, water or
roots to penetrate the soil. This type of soil structure
is very poor.

Green manuring promotes good soil structure because

organic matter and humus become glued between the clay
particles. The clay, organic matter and humus become
molded into aggregates which have spaces between them.
These aggregates fit together like a pile of pebbles,
rather than like tightly assembled bricks. Consequently,
air, water and roots can penetrate the spaces between
the aggregates.

Once again, it is the life of the soil which improves
soil fertility - in this instance soil structure. When
a green manure is turned under, the soil microorganisms
digest this organic matter and produce polysaccharides,
the glue which holds together clay, organic matter and
humus to form a soil aggregate.

I also wish to point out that this glue not only
improves the structure of clay soils, but also peat
soils. The structure of peat soils, with as much as
85 to 90 percent organic matter, can be greatly improved
by turning under a green manure crop. The fine particles of peat are bound together by the glue produced by
the soil organisms eating the green manure.

A good green manure crop can produce several thousands of pounds per acre of this glue. Soil aggregation
increases substantially when you turn under a green manure
crop. As the soil becomes more aggregated, it also
becomes looser. Cultivating becomes easier and does
not require as much energy. The plant roots do not
need to expend energy pushing through the soil. Water
penetrates into the soil, rather than puddling on the
surface. Air can now pass readily into the soil to
reach the soil organisms and plant roots. In addition,
while the green manure is growing, the plants protect
the aggregates from being shattered by rain. This can
be a significant benefit during the winter months.
/a8, p5, r3/

A big soil structure problem which green manuring
can help alleviate is hardpan - both surface and subsurface. Evaporation from the surface of an infertile,
bare soil draws up alkali salts. These salts form a crust

causing surface hardpan. Green manuring can help eliminate surface hardpan because the organic matter from green
manuring decreases evaporation from the soil surface and
prevents this alkali movement.

     For subsurface hardpan, deep rooted green manures,
like lupines, can penetrate hardpans and provide passages
for the roots of succeeding crops to follow. You might
think to use a chisel plow to break up subsurface hardpan. A chisel plow is a strong metal blade which slices
into the soil down as deep as 16 to 18 inches. This is
a good tool because it does not invert the soil, although
it does require a heavy machine and a great quantity of
energy to pull one through the soil. Try a deep rooted
green manure to break up hardpan first. Then use the
chisel plow. /g3, L6, p5/

Prevents Soil Erosion

     When water runs over a soil's surface, topsoil is
washed away with it. The soil erodes and is lost. The
more water your soil can absorb, the less your soil will
erode. One way to improve the water absorbing ability of
a soil is to add organic matter. Organic matter improves
the soil's water absorbing capacity because particles of
organic matter act like thousands of little sponges,
soaking up water. Besides sponge action, increased aggregation improves water retention. A well aggregated soil
has more spaces for water to penetrate and collect in, so
more water can infiltrate the soil. When you add organic
matter to the soil, the soil can absorb more water.
Green manuring increases soil organic matter content.

     There are other forms of erosion which green manure
protects the soil against. A green manure crop also
holds and protects the soil from wind, frost heaving and
raindrop splash. /a8, p5, w4/

     It is very important to add organic matter to the
soil so more water will penetrate the soil and erosion
will be prevented. Today, adding organic matter is·
especially important because soils low in organic matter
are even more easily eroded than those higher in organic

matter. As soils become more and more deficient in
organic matter, they erode away easier. To complicate
matters, soils where crops are grown in wide rows and
cultivated, like cotton, corn and potatoes, are highly
conducive to erosion. Soils which grow these crops can
definitely benefit from the addition of organic matter
through green manuring.

Erosion is a serious global problem. The earth's
supply of topsoil is being rapidly exhausted. An
average of 4 tons per acre of topsoil are washed off our
fields every year. In the United States, for example,
the average percentage of organic matter is at the low
level of 1.5%. The minimum recommended level is 5.0%
and the ideal level is 10.0% to 12.0%, as in virgin
topsoil. /a5, m3/

Certainly, animal manure is not the answer to
solving the erosion problem. There is just not enough
animal manure, sawdust, sludge and other such sources
of organic matter around to efficaciously supply our
soils with the quantity of organic matter which it needs
- even though these are good sources of organic matter.
We should look to green manuring as the answer to cur-
tailing erosion and revitalizing our soil. It is the
appropriate source of organic matter. If green manure
crops were grown and turned under regularly, the global
erosion problem could be alleviated.

Prevents Leaching of Mobile Nutrients

Mobile nutrients, like nitrogen and sulfur, are often
present in the soil solution when no crop is growing;
for example, during the fall and winter. If plant roots
are not growing and pumping up nutrients when mobile
nutrients are available in the soil solution, these
mobile nutrients can be leached out of the soil and lost.
Fortunately, a portion of these mobile nutrients are often
absorbed by fungi and bacteria.

When you grow a green manure, as a winter cover
crop, for example, you can prevent the loss of mobile
nutrients. The green manure incorporates mobile nutrients

into their plant parts, saves them and then returns them
to the soil when they decompose. If the soil was left
bare, the fall and winter rains would wash away the
mobile nutrients the green manure would save. This fact
is especially important for regions with mild, rainy
winters. In areas where the winters are cold and the soil
freezes, winter leaching is not as big a problem. However, growing a green manure is still beneficial to soil
fertility in cold climates. /m6, m7, m9, p5/

Soil Organisms Detoxify the Soil

     Green manuring also helps by detoxifying the soil.
While the soil organisms feed on organic matter, they
breakdown toxic substances - substances which are detrimental to both plants and people. Sodium is one such
substance. It is detrimental to plants and people when
absorbed in too large a quantity. Plants only need
minute quantities of sodium. When there is too much
sodium in the soil, the normal, healthy metabolism of
plants is interfered with. Think of it like drinking
salt water, Hylas. You just can't live on salt water.
It makes you thirstier than when you started drinking.

     Quite often, soils will inadvertantly get too much
sodium in them. Sometimes, there is too much sodium in
irrigation water or in soil ammendments. For example,
steer manure from feed lots often has too much sodium.
Or, just by being close enough to the sea coast, your
soil may have an inordinately large quantity of sodium.
Green manuring can help to minimize your difficulties
with sodium.

     This is how: When your soil is alive and feeding
voraciously on green manure, the microorganisms metabolize
sodium. They immobilize sodium by attaching it to larger
molecules. With sodium immobilized, its detrimental
impact on plants is greatly reduced. /c3/

     Other substances which are toxic to plants can also
be present in the soil on occasion, substances such as
herbicides. Sometimes, when herbicides have been used
to grow a crop one year, an herbicide residue will carry

over in the soil to the next year. If you change your
crop in that field, let's say from corn to soybeans, for
example, the soybean's growth will be hindered by the
herbicide. Green manuring can prevent this from occur-
ing in two ways. First, the soil creatures eat the
herbicides, breaking them up into molecules which are
not toxic to the following crop, soybeans in this case.
Second, the organic matter from green manuring absorbs
herbicides, thus removing them from the field of potent
action - the soil solution. /g5/

Retention of herbicides and other pesticides is
closely related to the soil organic matter content. The
more organic matter present, the greater quantity of
pesticide that can be absorbed. In some cases, growers
have added activated charcoal to the soil to soak up herbicide carryover. Organic matter from green manuring can
act as a carbon source to soak up pesticides and promote
microbial digestion which breaks down pesticides. /g4/

When crops are grown in live soils with plenty of
organic matter, the need to use pesticides for insect and
weed control becomes negligable. Green manuring prevents
pest problems so you can avoid using toxic substances.

Helps Control Insect Pests

Green manuring can help you avert insect pest
problems. This is primarily accomplished through rotation. By rotating your crops, you interfere with the
feeding and breeding grounds of insect pests. The
insect's host plant (your cash crop), which is normally
grown in a particular field, year after year, is no longer
there so the insect pests stop reproducing. Rotating
crops with a green manure can help control, for example,
corn insects in the Midwest, cotton insects in the
South and sugar beet nematodes in California. Instead
of growing the same crop year after year, rotate your
crops, preferably with a leguminous green manure. Your
insect pests will be reduced in number. This also holds
true for many plant diseases. /k3, s10/

I might add that turning under crop residues or

a green manure destroys the shelter, food supply and
breeding grounds of many insect pests, In particular,
corn borers and many types of cutworms are disturbed by
tillage. Preparing the soil to sow a green manure in fall
or tilling it under in spring, can have great benefits in
controlling these pests. /a8/

Extinguishes Weed Problems

Many green manures can smother out rankly growing
weeds. Perennial weeds are usually the culprits which
you have the most difficulty getting rid of. That nasty
perennial weed, Quackgrass (Agropyron repens), can be
controlled by growing two or three crops of buckwheat in
the summer. The buckwheat smothers it out. Then when you
turn it under, the decomposing action of the soil microorganisms destroys the rhizomes of Quackgrass. You
should follow the buckwheat with a fall sown crop of
Italian ryegrass. Turn it under in the spring.

Canada thistle (Cirsium arvense) is another noxious,
perennial weed which can be tamed by growing a green
manure, alfalfa. The combination of a strong stand of
alfalfa smothering out the thistle and mowing several
times during the growing season, starves the rhizomes
of this weed. A couple of seasons of this treatment,
greatly reduces the size of a Canada thistle infestation.
Even if you don't have a perennial weed problem, green
manure crops, like alfalfa, in your rotation can greatly
reduce the number of weeds in the following crop - corn,
for example. /L6/

You will be even more assured that the green manure
will outcompete the weeds if you optimize the soil
conditions favorable to the green manure plant. It
helps to have your soil analyzed before sowing in order
to see what your soil needs - add lime if it needs
calcium, for example. /m14/

Weed Competitors

Legumes	Non-legumes
Pigeon pea	Barley
Hyacinth bean	Sudangrass
Velvet bean	Pearl millet
Fava bean	Buckwheat
Cow pea	Rape
Alfalfa	Rye
	Sorghum
	Sunflower
	Corn

/a5, m14, m15, p5, s7/

Things to Watch Out For

It is important for you to be aware of some difficulties which may complicate green manuring. You can avoid problems while green manuring by choosing the right plant species for your special situation. Generally, there are certain green manure crops suited for any particular growing situation. The only time you cannot feed your soil with green manure is if you farm where there is not enough annual rainfall.

Moisture Requirements

As a general rule, a green manure should not be grown if the region's average, annual rainfall is <u>less than 16 inches</u> and irrigation water is not available. The reason for this is because with such little moisture, green manure does not increase the soil's humus or nitrogen content. Under arid conditions, the low moisture content of the surface soil, along with high temperature and a high degree of aeration, stimulates the direct oxidation – burning up – of organic matter, rather than humification. In regions with less than 16 inches of annual rainfall, a year of bare fallow – frequently cultivating the soil for an entire season every 3 to 4 years – is a more effective way of retaining the productive power of semi arid soils.

Bare fallow saves water and nutrients by not allowing weeds to grow. However, bare fallowing hastens the loss of humus. For areas with more than 16 inches of annual rainfall, green manuring is better than bare fallowing. With sufficient moisture, green manure is the best means to maintain and improve the soil's fertility.
/a3, a4, m1, m9/

Frost
        In citrus groves, there is some uncertainty whether
the presence of a winter cover crop increases the possibility of frost damage to the trees. Some growers will
not grow a winter cover crop under citrus trees due to
this possibility. They contend that bare soil stores heat
from the sun and keeps the temperature warmer in their
orchards.
        A study was conducted to examine this question:
Does the presence of a winter cover crop lower the air
temperature in an orchard, thus promoting frost? In
this study, it was found that with a cover crop the temperature near the soil surface was $1.0°F$ lower than without a cover crop. Five feet above the soil surface the
temperature was $0.1°F$ lower due to the presence of the
cover crop. Nevertheless, practically no difference was
found between the two regarding fruit production. From
these results, it appears that a winter cover crop increases the possibility of frost damage very little, if
at all. The benefits from green manuring with a winter
cover crop most likely far outweigh the small risk from
frost damage. /p5/

Overstimulation of Growth
        On rare occasions, legume green manures supply too
much nitrogen to orchard trees, causing the trees to produce too much wood and vegetation, rather than fruit.
This has occurred in New York peach orchards when clover
was used as a green manure and also in Oregon apple
orchards. Normally, green manure does not overstimulate
the growth of crops. /p5/

Heavy Metals
        If sludge which contained heavy metals, like Cad-
mium, has ever been added to your soil, you should be
careful. Green manuring can make heavy metals more
readily available for plants. You should beware of
soils which contain heavy metals which are toxic anyway.
As soil microrganisms digest organic matter, they excrete acids which improve the mobility of hazardous,

heavy metals. Green manuring may not be a good idea in soils which contain heavy metals. /c4/

Enemies of Green Manure

Like any plant, green manure crops have their insect and disease enemies. Fortunately, green manure crops are not often seriously affected by these enemies if properly rotated. When the same green manure is grown in one place repeatedly, a pest population may build up. In order to avoid the difficulties of a pest buildup on a green manure crop, it is important to rotate your green manures. Alternate a field with a new green manure crop every few seasons. An example of how you could rotate green manure crops in a California orchard is to grow the following legumes, each for two years: sour clover, purple vetch, Tangier pea, fava bean and fenugreek. /k3, p5/

Pests, Predators and Parasites

It is important to consider potential insect and disease pest problems when choosing a green manure crop. Some green manure crops are attacked by pests which also attack certain crops. For example, the cow pea is a valuable green manure crop for young pecan orchards, but when the trees begin producing nuts, cow peas can no longer be used as a green manure. The reason for this is because a squash bug attacks both the cow peas and the producing pecans. You are wise to examine pest lists for the green manure crops you wish to grow and your cash crop. This will help you avoid building up pest populations on the green manure crop. /p5, see y1/

Conversely, a green manure crop can be an aid to pest management. Green manure acts as a haven to beneficial insects - the predators and parasites. For example, green manures which produce nectar, like white clover, can boost predator and parasite populations by feeding these good guys. When you grow white clover in an orchard, their nectar can feed predators, like lacewings, or parasitic wasps, which help to keep aphid populations in check. /c3/

Planting a Green Manure

Before preparing the soil and sowing a green manure, it is best to have your soil analyzed in order to determine its fertility. Your soil analysis should include at least the following: 1) the percentage of organic matter, 2) levels of nitrogen, calcium, magnesium, potassium, phosphorus, sulfur and sodium, 3) the cation exchange capacity (CEC), 4) the percent base saturation of the cations: calcium, magnesium, potassium, hydrogen and sodium, and 5) the pH. Having your soil tested for its level of micronutrients, like zinc, copper, molybdenum, mangenese, iron, boron and chloride, can also be helpful when you have a soil problem not solved by macronutrient analysis and ammendment.

Unfortunately, some soils are naturally deficient in certain elements. If your soil is deficient in an important nutrient, your green manure may not grow well. Some soils are just not suited for a particular green manure species; the soil can be too alkaline (too high a pH), for example. Soil analysis helps you optimize the benefits of growing a green manure crop. By first analyzing your soil at a dependable laboratory and determining which nutrients your soil is deficient in, you can add these nutrients in a natural form before sowing the green manure. For example, some legumes, like red clover and sainfoin, require substantial quantities of calcium for sucessful growth. Lime (calcium carbonate) can be added to soils deficient in calcium to provide soil conditions suitable for growth of these legumes.

In general, it is helpful to know the nutritional requirements and conditions which promote the growth of

Natural Soil Ammendments	Major Elements
Wood ashes	potassium
Granite dust	potassium
Green sand	potassium
Soft rock phosphate	phosphorus, calcium
Bone meal	phosphorus, calcium
Gypsum	sulfur, calcium
Dolomite	magnesium, calcium
Lime	calcium
Kelp	trace elements

the legume-Rhizobium symbiosis. Here are a few guide-
lines for optimizing nitrogen fixation and the growth of
legumes for green manure:

1. Adequate levels of phosphorus, potassium
    and sulfur are required for normal growth.
2. The micronutrients - molybdenum, boron,
    cobalt, vanadium and iron play specific
    roles in the fixation process.
3. Chemical nitrogen fertilizer tends to
    reduce both nodule formation and nitro-
    gen fixation.
4. The symbiosis is sensitive to exceedingly
    acid soil. However, some legumes tolerate
    a lower pH than others.
5. Calcium promotes fixation.
    /b3, see w6/

It is important to become familiar with the growth
requirements of individual green manure species. This is
why I will give you a book of species descriptions, Hylas.
Know each species individually so you can give it proper
care. Know whether it will grow in your soil and climate.

Once you know your soil and have chosen the appropriate green manure species for your soil and climate,
you are ready to plant. The steps to planting a green
manure crop are:

1. Irrigate, if available and necessary.

2. When sufficiently dried, thoroughly <u>cultivate</u>
    the soil. Prepare the soil to the degree of
    fineness and firmness required by the green
    manure being sown. Some green manures, like
    clovers, need fine, firm soil to germinate in.
    While you are cultivating is a good time to
    work in the needed soil ammendments, as deter-
    mined by your soil analysis.

3. <u>Drill or broadcast the seed</u>. In general, the
    larger a green manure seed, the deeper it can
    be sown, or the smaller the seed, the closer to
    the surface it should be placed. Another impor-
    tant factor to consider when sowing is how fast
    the soil dries out. Sandy topsoil dries out
    faster than clay. Therefore, in sand you should
    sow your seed a little deeper.

    *Green manure seed usually germinates better when
    drilled than broadcast because all of the seed is
    placed at the right soil depth. Additionally,
    the green manure plants are evenly spaced, grow-
    ing in rows when drilled. These days, it is
    even possible to drill seed into sod. This is
    accomplished with a machine known as a sod seed-
    er. A sod seeder slices a slit in the sod before
    placing the seed in the soil.

    *Broadcasting seed - scattering seed on the soil
    surface - is suitable for many green manure
    crops, especially those which germinate rapidly.
    You can broadcast seed from an airplane, tractor
    or by walking in the field yourself. Broad-
    casting equipment which spins the seed from a
    wheel onto the soil provides much more even
    coverage than when you sow by tossing it out of
    your hand.

    *For some green manure crops, especially small
    seeded legumes which germinate slowly, like the
    clovers, you may want to sow a nurse crop before,
    or along with the legume. A nurse crop, such as

rye or barley, is grown to protect the germinating legume by catching rain and irrigation water
before it strikes the small germinating legume.
Nurse crops also prevent weeds from outcompeting
the slow starting legumes.
*The best nurse crop is barley because it does not
grow as tall as the other grains and it occupies
the soil for a shorter period of time. The
other three nurse crops, in order of preference,
are: rye, wheat and oats. Oats is the least
desirable nurse crop, particularly for clover,
because they are leafy and their shade impedes
the clover. Oats often lodge. They also draw
moisture more heavily than the other nurse
crops. If you want to use oats as a nurse crop,
you can sow the oat seed thinly (1 bushel/acre).
4. <u>Cover the seed</u> by harrowing or discing lightly.
5. <u>Irrigate</u> if possible.
    /a7, b1, see f2, m9, r1/
When you sow legumes, it is sometimes necessary to
inoculate the seed with the appropriate Rhizobium bacteria species. Inoculation is an inexpensive and simple
process. You just lightly moisten your legume seed with
water so the bacteria inoculum will stick. Then pour in
the bacteria granules with the seed and mix it thoroughly with the seed. It is important to sow your seed
immediately after inoculation.

Pelletized legume seed is becoming more commonly
available. This is seed which has been coated with
ground lime, gum arabic, as an adhesive, and the bacterial
inoculum. Coating seed in this manner has proven to
increase nodulation of legume roots. /h7, h8/

When you are growing a legume on land which has
never grown the crop before, you should inoculate the
seed. If you are certain that the soil carries the
proper Rhizobium species, you can skip inoculation. In
some regions the soil is already naturally inoculated for
particular legumes. For example, Florida beggarweed,

cow peas, velvet beans, Crotolaria, Lespedeza, alyceclover, hairy indigo and common sesbania are ususally
naturally inoculated in regions to which they are adapted.
In the Western half of the United States, alfalfa and
sweet clover are generally naturally inoculated, but
sometimes need inoculation when grown for the first time.
Soybeans should always be inoculated when grown for the
first time. Clover seldom requires inoculation because
of the wide distribution of wild and cultivated clover.
/m6/
     There are different Rhizobium species which effect-
ively inoculate their own particular legume group. These
Rhizobium species do not cross-inoculate between groups
so it is important to use the correct species. You can
be sure by checking the following chart.

          Legume Cross-Inoculation Groups
          and Their Rhizobium Species
1. Alfalfa group - Rhizobium meliloti
        Alfalfa (Medicago sativa)
        Bur clover (Medicago spp.)
        Fenugreek (Trigonella foenum-graecum)
        Sweet clover (Melilotus spp.)
        Yellow trefoil (Medicago lupulina)
2. Clover group - Rhizobium trifolii
        Clovers (Trifolium spp.)
3. Vetch group - Rhizobium leguminosorum
        Horse bean (Vicia faba)
        Lentil (Lens esculenta)
        Peas (Pisum spp.)
        Tangier pea (Lathyrus tingitanus)
        Vetches (Vicia spp.)
4. Bean group - Rhizobium phaseoli
        Beans (Phaeolus spp.)
5. Lupine group - Rhizobium lupini
        Lupines (Lupinus spp.)
        Serradella (Ornithopus sativus)

Legume Cross-Inoculation Groups and Their Rhizobium Species

6. Soybean group - <u>Rhizobium japonicum</u>
   Soybean (<u>Glycine max</u>)
7. Cowpea group - *
   Adzuki bean (<u>Phaseolus angularis</u>)
   Cow pea (<u>Vigna sinensis</u>)
   Crotolaria (<u>Crotolaria</u> spp.)
   Florida beggarweed (<u>Desmodium purpureum</u>)
   Hyacinth bean (<u>Dolichos lablab</u>)
   Jack bean (<u>Canavalia ensiformis</u>)
   Kudzu (<u>Pueraria hirsuta</u>)
   Lespedeza (<u>Lespedeza</u> spp.)
   Lima bean (<u>Phaseolus limensis</u>)
   Moth bean (<u>Phaseolus aconitifolius</u>)
   Mung bean (<u>Phaseolus aureus</u>)
   Peanut (<u>Arachis hypogaea</u>)
   Pigeon pea (<u>Cajanus cajan</u>)
   Sword bean (<u>Canavalia gladiata</u>)
   Tepary bean (<u>Phaseolus acutifolius</u>)
   Velvet bean (<u>Stizolobium deeringianum</u>)
   * This group has not attained species
     status .
   /a6, b3, p5, t1/

For green manures which are not legumes, like barley, rye, oats, wheat, corn and sunflowers, you can increase their growth by inoculating your seed with a free living, nitrogen-fixing bacteria named Azotobactor. Azotobactor fixes nitrogen from the air and produces hormones which stimulates crop growth. The crop roots in turn exude sugars which feed the free living bacteria so it can multiply and increasingly benefit the crop. The nutritional requirements and growing conditions needed by Azotobactor are similar to those of the legume-Rhizobium symbiosis. However, Azotobactor also benefits from a good supply of organic matter in the soil and a high pH - an optimum between 7.2 and 7.6. Below a pH of 6.5, its growth is severely limited. Like Rhizobium, Azotobactor's ability to fix nitrogen is inhibited by chemical nitrogen. Additionally, soils with an excess of phosphate, such as those overfertilized with superphosphate, can inhibit Azotobactor.

The free living bacteria, Azotobactor, can help a green manure meet its phosphorus requirement. It does this in two ways. First, it produces vitamins which benefit the growth of mycorrhizal fungi. These fungi act like root hairs and aid the plant roots in obtaining phosphorus. Second, Azotobactor absorbs phosphates from the soil which are otherwise bound into forms which plants can't use. Phosphate released upon the death of Azotobactor is in a form usable by plants.

The life span of an individual Azotobactor bacterium is 24 hours. Provided with the right soil conditions, Azotobactor, like other bacteria, can reproduce rapidly. The greater number of Azotobactor working in your soil, the more nitrogen will be fixed for your green manure crop. Overall, green manuring with a non-legume can be enhanced by inoculating with Azotobactor. /c2/

Another aspect to consider while sowing a green manure crop is that some legumes have hard seed that will not germinate readily without being scarified (scratched to make the seed coat thinner). Legumes, like sweet clover and Lespedeza, have a high percentage of hard seed. Hard seed should be scarified if prompt germination is desired. Otherwise, you can sow hard seed in the fall and, by spring, a larger percentage will be in a germin-able condition. Sometimes, you can buy seed which has already been scarified. If it hasn't, here is one way to scarify it yourself: Place your seed and some sand in a sack with a tight mesh so the seed-sand mixture won't fall out. Then drive over the sack in a car several times. /m6/

Most green manure seed does not need to be scarified. Some seed, you don't even need to buy once it has initial-ly been sown in your field. Certain winter legumes, if

allowed to produce seed, will volunteer - reproduce themselves enough without sowing - for the following season.
Legumes which successfully volunteer are the least
expensive to use for green manuring. Winter legume
volunteers can be divided into two classes: 1) those
which mature early in the spring, before it is time to
sow the summer crop, and 2) those which mature seed late.
By allowing occasional seed crops, every 4 to 5 years for
the late maturing green manures, you will insure volun-
teering. Bur clover can be successfully handled in this
manner, allowing it to mature once every 5 years. Another
method to volunteer bur clover is to leave strips of it
in the field you are going to sow summer crops in. Then
turn the bur clover under after the seed has matured.

Volunteers

Beggarweed	Desmodium purpureum
Black medic	Medicago lupulina
Bur clover	Medicago spp.
Crimson clover	Trifolium incarnatum
Annual Lespedezas	Lespedeza stipulacea
	L. striata
Persian clover	Trifolium resupinatum
Rose clover	Trifolium hirtum
Rough pea	Lathyrus hirsutus
Sour clover	Melilotus indica

Turning Under a Green Manure

When you turn under a green manure, you want to do it in a manner which most effectively feeds the soil organisms. In other words, you want to optimize decomposition, humification and the feeding of the plants by the growing, reproducing and dying soil organisms. As I told you a year ago Hylas, soil organism growth is promoted by the proper levels of moisture, air (particularly oxygen) and temperature. By considering and working with these factors, you can optimize the growth of soil organisms and the effectiveness of your green manure in feeding your crops. Remember, these three factors work together. They compliment each other. /m6, p5, r3, w4/

Sixty to eighty percent of the water holding capacity of a soil is optimum for decomposition. You do not want too much water in the soil, or air will not be able to penetrate. This sort of situation occurs in flooded rice fields. In soggy soil, air and oxygen cannot penetrate so decomposition proceeds slowly under anaerobic conditions. Too much water is not only troublesome because it causes slow decomposition, but it also promotes denitrification - the loss of soil nitrogen by vaporizing back into the atmosphere. You want the soil to be moist, but not soggy, or aerobic decomposition will be stifled.

A green manure crop decomposes most readily in moist soil during hot weather. As the temperature declines in the fall or rises in the spring, you can expect a corresponding decrease or increase in the activity of soil life. In warm, moist, well aerated soil, you can expect almost complete decomposition to take place in 6 weeks.

Decomposition of green manure goes on more rapidly
in light soils than heavy ones. This is due to aeration.
This ease of aeration in light soils also allows them
to dry out quickly. On sandy soils, green manure should
be cultivated under more deeply. This way the decomposing
organic matter will not dry out so readily and will
continue to decompose after the first 3 to 4 inches of
light soil has dried out. Likewise, if you are growing
a cash crop which requires a great deal of cultivation for
weed control, it is advisable to turn under the green
manure somewhat in excess of the depth of cultivation.
Cultivation causes organic matter to decompose too
quickly because it aerates the soil and aids the sun in
increasing the soil temperature. /m9, p5/

From what I've said about the depth of turning a
green manure, you would think I am saying you should
turn under a green manure with a plow. You should,
but only sometimes. The time to use a plow is when your
topsoil is very deep, let's say 12 inches for example, as
may occur in peaty soil; or when your soil has no topsoil. If your soil has a thin or moderate layer of
topsoil stratified over a mineral subsoil, you do not want
to invert your soil with a plow; thus placing the topsoil under a layer of subsoil. Instead, you should
cultivate under a green manure crop with a <u>disc</u>, or,
even better, a shallow rototiller. You can thoroughly
mix the green manure with the topsoil by using a disc.
By incorporating a green manure into a soil with a disc,
rather than a plow, you can keep the soil stratified -
topsoil over subsoil - as it naturally occurs. Also,
you can avoid plowing up weed seeds. /m9, r2/

Sometimes, a green manure may be too stringy, or too
tall, to be easily turned under with a disc, or even a
plow. This situation often occurs with vetch or field
peas. There are two ways to deal with these stringy
green manures before discing. You can either mash down
the vines with a roller, or you can mow first, before
tilling them under. Rolling or mowing makes tillage of

stringy green manures much easier.

You could decide to leave your green manure on the
soil surface as a mulch. This is helpful in preventing
the soil from drying out. However, a green manure completely turned under will decompose more rapidly and
feed the succeeding crop faster than one left on the
surface or partially turned under. Also, crops directly
seeded may have difficulty in germinating through a mulch.
One situation where leaving a green manure mulch on the
soil surface may be beneficial is when you are transplanting a cash crop in a dry region.

Regarding the green manure itself, it is important
to cultivate the crop into the soil while it is still
green. Decomposition is retarded when you allow a green
manure to dry out before turning it under. The soil
organisms cannot digest dry vegetation. It must be moist.
The internal moisture of the vegetation greatly aids
decomposition. This is why it is called green manure
instead of brown manure.

In addition, the rate of decomposition by soil
organisms is hastened by an ample supply of nitrogen in
the soil. Remember, soil organisms need nitrogen just
like plants do. As plants grow older, they contain a
larger percentage of carbon and a smaller percentage of
nitrogen - a wide C/N ratio. The younger the green
manure, the larger the percentage of nitrogen - the
narrower the C/N ratio. Young, succulent green manure
decomposes much more readily than older or mature plants.
Similarly, legumes decompose more rapidly than non legumes
for the same reason - the legumes have a narrower C/N
ratio.

The best time to plow under a leguminous green manure
is when the plants are flowering. This is the time when
the legume-Rhizobium symbiosis has fixed the largest
quantity of nitrogen from the air. Also, the legume is
still succulent for easy digestion by the soil organisms.
If you allow the legumes to produce seed, most of the
nitrogen assimilated during growth is located in the

seeds. Consequently, the nitrogen is not available to the soil organisms and succeeding crops, but locked up in the seeds. /c1, p5/

Turn under non legume green manures, like buckwheat and grasses while they are young, before they flower and set seed. This way they will have a narrower C/N ratio than when they get older. Also, unlike older vegetation, the decomposing, young grasses will not steal nitrogen from the soil which would otherwise go to the growing plants.

Another way to improve the decomposition of green manure is by inoculating the soil with beneficial microorganisms when cultivating the green manure under. This can be accomplished through several different materials - by adding stable manure, compost or biodynamic field sprays. These materials add more of the good guys, fresh decomposers, to the soil to actively initiate digestion of the green manure. /m16, p5/

Nonetheless, it is important to recognize that these beneficial decomposers can attack germinating seeds if you sow your cash crop too soon after turning under a green manure. Green manure crops must be turned down a sufficient length of time - about 3 weeks - before an annual crop is planted. This way, the organic matter can be digested by the first few successional groups of decomposers before sowing. If a green manure is turned under and a crop sown too soon afterwards, these initial decomposers will attack the germinating seeds. Fungi from the actively decomposing green manure can invade the germinating seedlings.

It may seem strange, but oil seeds are more readily damaged by the decomposers of freshly turned under green manure, than starchy seeds. (But remember, starchy seeds are also attacked.) The sunflower is one exception to this rule. Sunflower seeds are not as susceptible as other oil seeds to damage by the early invaders of fresh green manure. The quick germination of sunflowers may explain their resistance.

There is a difference between oily and starchy seeds
which may explain their varying suseptibility to damage.
Seeds rich in oil require more oxygen for germination than
starch seeds. Maybe there is just not enough oxygen in the
soil and the oil seeds are overtaken by the decomposers?

Oil seeds	Starch seeds
cotton	corn
soybeans	oats
flax	wheat
peanuts	buckwheat
mustard	
clover	

The damage to oil seeds from green manure is con-
fined largely to the first stages of decomposition. Once
decomposition has progressed for a few weeks, the seeds
are no longer in danger from being attacked by the
decomposers of green manure. Therefore, to prevent damage
and insure good germination, allow two weeks to elapse
before sowing starchy seeds and three weeks before
sowing oily seeds. /f1, m6, p5, w4/
One rule you should always try to observe is:
Never allow the soil to be bare for more than three
weeks, especially in sandy or porous soils. If you do,
rain can more easily leach away mobile soil nutrients,
like nitrogen and sulfur. In sands or porous soils,
leaching is rapid. While in heavy clays, leaching
occurs more slowly. The only exception to this rule
is in the North, where the soil freezes during the winter.
Frozen soil prevents nutrients from washing away. On the
other hand, in the South and other areas with mild, rainy
winters, mobile soil nutrients can be easily leached out
of the soil. If your farm is in a mild climate, sow rye
or some other winter growing crop after your fall harvest
or after you turn under your summer green manure. This
winter crop will prevent leaching of mobile nutrients
released by the decaying green manure. /m6, w4/

"That's it Hylas. I've just told you everything I know about green manure." Hylas was drowsy. The sun was setting. He had listened attentively to Eatmore, but had reached his saturation threshold. His little rabbit brain couldn't absorb any more information today.

"Now Hylas, at this point, I wish to give you a book. It is a very valuable source of information which can be found on any earthworm's bookshelf. This book is passed on from generation to generation of earthworm. Only earthworms know of this book. You are the first non-earthworm to hold it in your paws. I expect you to value this book as an earthworm would value it. The print is rather small, but you can use a magnifying glass to read it."

Hylas woke up upon the presentation of such a gift. He was honored. The bestowing of the earthworms' book shook him to the core. Hylas was Eatmore's student. True, he only received instruction for two full days, yet relative to the lifespan of an earthworm - 2 years - this was a long time. Eatmore passed on knowledge and responsibility to his one student, Hylas. Hylas' face beamed with gratitude. Eatmore felt this. No words were needed.

Hylas' eyes were watery as he pulled a magnifying glass out of his pocket and began reading.

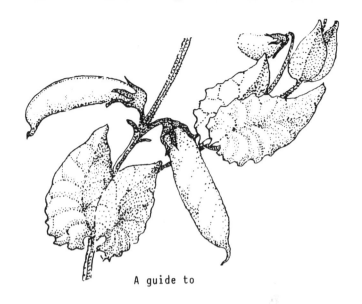

A guide to

GREEN MANURE,

COVER CROPS

AND

CULTIVATED LEGUMES

by,

Ralph Waldo Earthworm

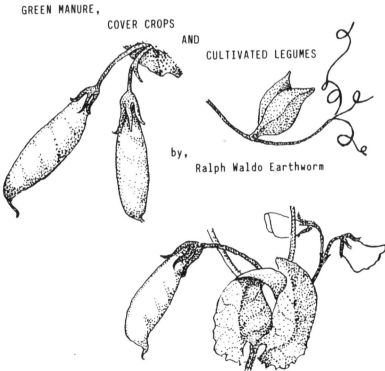

⸎ How to Choose Plants for Green Manuring ⸎

   This book is a guide. It will help you discover
which soil improving plants are best adapted to your
farm. Primarily, this book will help you choose appropriate green manure species to feed the soil and cover
crop species to prevent erosion. However, many other
plant species have been included because they can also
improve your soil when used as cash crops in rotations
and as pasture plants.

   In the front section, you will find the Soil-Climate
Key. This Key is a shortcut. By using it, you will save
time. You won't have to read about all of the 127 species
in order to decide which green manures you should grow.
To use the Key, you first need to know your soil type.
The Key lists five soil types to choose from: Sandy,
Sandy loam, Loam, Clay loam and Clay. After you have
found your plants under one of these soil types, make
sure you turn to the "Many Soils" section. This section
will help you increase your list of species. These
plants are adapted to a wider range of conditions. There
are also three additional categories for extremely Acid,
Alkaline or Saline soil types.

   Most of us have visited the beach so we can distinguish sand from the other soil types. A loam soil is one
which is rich in organic matter — this is what you are
working toward. You can tell if a soil has clay in it
by squeezing chunks of earth between your thumb and
forefinger. If it is pliable, like bread dough, then it
contains clay. Another way to distinguish sand from
clay is with a microscope. Through a microscope, you
can see that sand particles are much larger than clay
particles.

Next, you choose your climate. We all have a
general feeling for the difference between a cool or a
hot climate, and a moist or dry one. The South would be
considered to be cool during the winter and the extreme
northerly latitudes and high altitudes would be cool in
the summer. For the most part, these climate catagories
are general guides to steer you in the right direction
when choosing your green manure species. As you read the
species descriptions, you will discover more specific
details about the climatic conditions required by each
plant.

Behind each species, you will find an A, B, or P.
They stand for Annual, Biennial and Perennial. Also,
some species will have an N behind them. This means
that the plant is a Non legume. You will find that most
of the plants are legumes.

From using these categories, you can make a list of
plants suitable for your own growing situation. After
you have obtained your list of species adapted to your
farm, turn to each individual species in the text. They
are listed alphabetically by their latin (scientific)
names.

FOR EXAMPLE: Let's say, you wanted to grow a winter
cover crop between summer cash crops. Your soil type was
a sandy loam and winter climate was cool and moist.
Because you were going to follow the winter cover crop
with a row crop, you wanted to grow annual, green manure
species to be turned under in the spring. Your objective
in growing the winter cover crop was to hold the soil,
thus preventing erosion, and to add organic matter and
nitrogen, in order to improve soil fertility for the
following crop.

You first turn to SANDY LOAM, then under MANY
CLIMATES you find Italian Ryegrass, your first possibility. Moving along, you go to COOL CLIMATE, in conjunc-
tion with your winter climatic conditions. There, you

find seven annual legumes which you may possibly use in
your winter cover crop mixture. You now have eight
annual plant species that would be suitable for growth
on sandy loam soil, in a cool, moist climate. They are:

> Rough Pea (Lathyrus hirsutus)
> Tangier Pea (Lathyrus tinginatus)
> Italian Ryegrass (Lolium multiflorum)
> Large White Lupine (Lupinus albus)
> Yellow Lupine (Lupinus luteus var. sativa)
> Field Pea (Pisum arvense)
> Common Vetch (Vicia sativa)
> Hairy Vetch (Vicia villosa)

Next, turn to the MANY SOILS section to increase
your list.

Now, using the latin names, you turn to each individual species (listed alphabetically in the following text)
to get a more detailed description of the optimum growing
conditions for each plant. From the descriptions, you
choose the species best suited to your growing conditions.
It is advisable to grow several green manure species
your first year, in order to test them out. And, while
you are at it, grow some green manure for seed. As
interest and the need for green manure increases, having
more seed available will be important. For those of you
who are really motivated, you could turn your seed
production into a regional breeding program.

~ The Soil-Climate Key ~

SANDY SOIL

MANY CLIMATES: Rye (Secale cereale)   A, N

COOL CLIMATE: California Bromegrass (Bromus carinatus) A, N
Narrow-leaved Lupine (Lupinus angusti-folius)   A
Yellow Lupine (Lupinus luteus var. sativus)   A
Serradella (Ornithopus sativus)   A
Monantha Vetch (Vicia monanthos)   A
Hairy Vetch (Vicia villosa)   A
Spurry (Spergula arvensis)   A, N

COOL AND
DRY CLIMATE: Kidney Vetch (Anthyllis vulneraria)   P
Cicer Milkvetch (Astragalus cicer)   P

HOT CLIMATE: Peanut (Arachis hypogaea)   A
Sunn Hemp (Crotolaria juncea)   A
Crotolaria (Crotolaria spectabilis)   A
Dalea (Dalea alopecuroides)   A
Hairy Indigo (Indigofera hirsuta)   A
Tepary Bean (Phaseolus acutifolius)   A
Velvet Bean (Stizolobium deeringianum)   A

HOT AND
DRY CLIMATE: Moth Bean (Phaseolus aconitifolius)   A

SANDY LOAM SOIL

Alyceclover (Alysicarpus vaginalis) A

MANY CLIMATES: Italian Ryegrass (Lolium multiflorum) A,N

COOL CLIMATE: Turnip (Brassica rapa) B, N
Smooth Bromegrass (Bromus inermis) P, N
Rough Pea (Lathyrus hirsutus) A
Grass Pea (Lathyrus sativus) A
Tangier Pea (Lathyrus tinginatus) A
Large White Lupine (Lupinus albus) A
Yellow Lupine (Lupinus luteus var.
sativus) A
Egyptian Lupine (Lupinus termis)
Serradella (Ornithopus sativus) A
Field Pea (Pisum arvense) A
Common Vetch (Vicia sativa) A
Hairy Vetch (Vicia villosa) A

HOT CLIMATE: Peanut (Arachis hypogaea) A
Crotolaria (Crotolaria spectabilis) A
Beggarweed (Desmodium purpureum) A
Hyacinth Bean (Dolichos lablab) A
Hairy Indigo (Indigofera hirsuta) A
Foxtail Millet (Setaria italica) A, N
Sorghum (Sorghum bicolor) A, N
Cow Pea (Vigna sinensis) A

HOT AND
DRY CLIMATE: Moth Bean (Phaseolus aconitifolius) A

LOAM SOIL

Ball Clover (Trifolium nigrescens)

MANY CLIMATES: Redtop (Agrostis alba) P, N
Italian Ryegrass (Lolium multiflorum) A,N
Alfalfa (Medicago sativa) P
Yellow Sweet Clover (Melilotus offici-
nalis) B
Rye (Secale cereale) A, N

COOL CLIMATE: Rape (Brassica napus) A or B, N
Chess (Bromus tectorum) A, N
Rough Pea (Lathyrus hirsutus) A
Grass Pea (Lathyrus sativus) A
Tangier Pea (Lathyrus tinginatus) A
Toothed Bur Clover (Medicago hispida) A
Black Medic (Medicago lupulina) A or B
Sour Clover (Melilotus indica) B
Proso Millet (Panicum miliaceum) A, N
Timothy (Phleum pratense) P, N
Field Pea (Pisum arvense) A
Canada Bluegrass (Poa compressa) P, N
Kentucky Bluegrass (Poa pratensis) P, N
Crimson Clover (Trifolium incarnatum) A
White Clover (Trifolium repens) P
Subterranean Clover (Trifolium subterra-
nean) A
Fenugreek (Trigonella foenum-graecum) A
Winter Wheat (Triticum aestivum) A, N
Purple Vetch (Vicia atropurpurea) A
Common Vetch (Vicia sativa) A
Narrowleaf Vetch (Vicia sativa var. nigra)
A

COOL AND
DRY CLIMATE: Barley (Hordeum vulgare) A, N

LOAM SOIL

HOT CLIMATE: Orchardgrass (<u>Dactylis</u> <u>glomerata</u>) P, N
                  Sericea Lespedeza (<u>Lespedeza</u> <u>cuneata</u>) P
                  Korean Lespedeza (<u>Lespedeza</u> <u>stipulacea</u>) A
                  Common Lespedeza (<u>Lespedeza</u> <u>striata</u>) A
                  Texas Millet (<u>Panicum</u> <u>texanum</u>) A
                  Pearl Millet (<u>Pennisetum</u> <u>typhoides</u>) A, N
                  Kidney Bean (<u>Phaseolus</u> <u>vulgaris</u>) A
                  Kudzu (<u>Pueraria</u> <u>hirsuta</u>) P
                  Sudangrass (<u>Sorghum</u> <u>bicolor</u> var. sudanen-
                  se) A, N
                  Velvet Bean (<u>Stizolobium</u> <u>deeringianum</u>) A

HOT AND
DRY CLIMATE: Sesbania (<u>Sesbania</u> <u>macrocarpa</u>) A

CLAY LOAM SOIL

Ball Clover (Trifolium nigrescens)

MANY CLIMATES: Redtop (Agrostis alba) P, N
White Sweet Clover (Melilotus alba) B
Rye (Secale cereale) A, N

COOL CLIMATE: Smooth Bromegrass (Bromus inermis) P, N
Grass Pea (Lathyrus sativus) A
Tangier Pea (Lathyrus tinginatus) A
Toothed Bur Clover (Medicago hispida) A
Timothy (Phleum pratense) P, N
Field Pea (Pisum arvense) A
Canada Bluegrass (Poa compressa) P, N
Kentucky Bluegrass (Poa pratensis) P, N
Red Clover (Trifolium pratense) B or
short-lived P
White Clover (Trifolium repens) P
Purple Vetch (Vicia atropurpurea) A
Horse Bean (Vicia faba) A
Hungarian Vetch (Vicia pannonica) A

HOT CLIMATE: Orchardgrass (Dactylis glomerata) P, N
Texas Millet (Panicum texanum) A
Kudzu (Pueraria hirsuta) P
Sudangrass (Sorghum bicolor var. sudanen-
se) A, N

CLAY SOIL

Ball Clover (Trifolium nigrescens)

MANY CLIMATES: Tall Fescue (Festuca arundinacea) P, N
Rye (Secale cereale) A, N

COOL CLIMATE: Small Blue Lupine (Lupinus angustifolius
var. caeruleus) A
Perennial Lupine (Lupinus perennis) P
Toothed Bur Clover (Medicago hispida) A
Timothy (Phleum pratense) P, N
Canada Bluegrass (Poa compressa) P, N
Red Clover (Trifolium pratense) A or B
White Clover (Trifolium repens) P
Ladino Clover (Trifolium repens var.
latum) P
Persian Clover (Trifolium resupinatum) A
Hungarian Vetch (Vicia pannonica) A

HOT CLIMATE: Bermudagrass (Cynodon dactylon) P, N
Orchardgrass (Dactylis glomerata) P, N
Sericea Lespedeza (Lespedeza cuneata) P
Texas Millet (Panicum texanum) A, N
Kudzu (Pueraria hirsuta) P

MANY SOILS

MANY CLIMATES: Redtop (Agrostis alba) P, N
Oats (Avena sativa) A, N
Tall Fescue (Festuca arundinacea) P, N
Italian Ryegrass (Lolium multiflorum) A,N
Bird'sfoot Trefoil (Lotus corniculatus,
L. tenuis) P
Black Medic (Medicago lupulina) A or B
Alfalfa (Medicago sativa) P
Sweet Clovers (Melilotus alba, M. offi-
cinalis) B
Rye (Secale cereale) A, N
Rose Clover (Trifolium hirtum) A

COOL CLIMATE: Scotch Kale (Brassica oleracea) A, N
Field Bromegrass (Bromus arvensis) A, N
Rescuegrass (Bromus catharticus) P, N
Smooth Bromegrass (Bromus inermis) P, N
Rough Pea (Lathyrus hirsutus) A
Grass Pea (Lathyrus sativus) A
Tangier Pea (Lathyrus tinginatus) A
Lentil (Lens esculenta) A
Perennial Ryegrass (Lolium perenne) P, N
Large White Lupine (Lupinus albus) A
Spotted Bur Clover (Medicago arabica) A
Toothed Bur Clover (Medicago hispida) A
Button Medic (Medicago orbicularis) A
Sour Clover (Melilotus indica) B
Timothy (Phleum pratense) P, N
Field Pea (Pisum arvense) A
Hop Clovers (Trifolium agrarium, T. pro-
cumbens) A
Alsike Clover (Trifolium hybridum) P
Crimson Clover (Trifolium incarnatum) A
Red Clover (Trifolium pratense) B or
short-lived P

MANY SOILS

COOL CLIMATE: White Clover (Trifolium repens) P
                Ladino Clover (Trifolium repens var.
                latum) P
                Arrowleaf Clover (Trifolium vesiculosum)
                A
                Bird Vetch (Vicia cracca) P
                Woollypod Vetch (Vicia dasycarpa) A
                Horse Bean (Vicia faba) A
                Common Vetch (Vicia sativa) A
                Hairy Vetch (Vicia villosa) A

COOL AND
DRY CLIMATE: Crested Wheatgrass (Agropyron desertorum)
                P, N
                Smallhop Clover (Trifolium dubium) A

HOT CLIMATE: Peanut (Arachis hypogaea) A
                Jack Bean (Canavalia ensiformis) A
                Sword Bean (Canavalia gladiata, C. obtus-
                ifolia) A
                Guar (Cyamopsis tetragonoloba) A
                Orchardgrass (Dactylis glomerata) P, N
                Dalea (Dalea alopecuroides) A
                Hyacinth Bean (Dolichos lablab) A or
                semi P
                Buckwheat (Fagopyrum esculentum) A, N
                Soybean (Glycine max) A
                Sunflower (Helianthus annuus) A
                Korean Lespedeza (Lespedeza stipulacea) A
                Common Lespedeza (Lespedeza striata) A
                Proso Millet (Panicum miliaceum) A, N
                Pearl Millet (Pennisetum typhoides) A, N
                Adzuki Bean (Phaseolus angularis) A
                Mung Bean (Phaseolus aureus) A
                Lima Bean (Phaseolus limensis) A
                Kidney Bean (Phaseolus vulgaris) A

MANY SOILS

HOT CLIMATE: Kudzu (<u>Pueraria hirsuta</u>) P
Sorghum (<u>Sorghum bicolor</u>) A, N
Sudangrass (<u>Sorghum bicolor</u> var. sudan-
ense) A, N
Velvet Bean (<u>Stizolobium deeringianum</u>) A
Strawberry Clover (<u>Trifolium fragiferum</u>)
P
Cow Pea (<u>Vigna sinensis</u>) A
Corn (<u>Zea mays</u>) A, N

HOT AND
DRY CLIMATE: Moth Bean (<u>Phaseolus aconitifolius</u>) A
Sesbania (<u>Sesbania macrocarpa</u>) A

ACID SOIL

Redtop (Agrostis alba) P, N
Jack Bean (Canavalia ensiformis) A
Sword Bean (Canavalia gladiata, C. obtus-
ifolia) A
Crotolaria (Crotolaria spectabilis) A
Dalea (Dalea alopecuroides) A
Beggarweed (Desmodium purpureum) A
Hyacinth Bean (Dolichos lablab) A
Buckwheat (Fagopyrum esculentum) A, N
Sericea Lespedeza (Lespedeza cuneata) P
Korean Lespedeza (Lespedeza stipulacea) A
Common Lespedza (Lespedeza striata) A
Marsh Bird'sfoot Trefoil (Lotus pendun-
culatus) P
Large White Lupine (Lupinus albus) A
Yellow Lupine (Lupinus luteus var.
sativus) A
Rye (Secale cereale) A, N
Subterranean Clover (Trifolium subterra-
nean) A

CALCAREOUS OR
ALKALINE SOIL

Kidney Vetch (Anthyllis vulneraria) P
Rape (Brassica napus) A or B, N
Turnip (Brassica rapa) B, N
Barley (Hordeum vulgare) A, N
Large Blue Lupine (Lupinus pilosus var.
var. caruleus) A
Black Medic (Medicago lupulina) A or B
Sainfoin (Onobrychis sativa) P
Kidney Bean (Phaseolus vulgaris) A
Field Pea (Pisum arvense) A
Berseem Clover (Trifolium alexandrinum) A

SALINE SOIL

Rape (Brassica napus) A or B, N
Rescuegrass (Bromus catharticus) P, N
Tall Fescue (Festuca arundinacea) P, N
Jack Bean (Canavalia ensiformis) A
Sword Bean (Canavalia gladiata, C. obtus-
ifolia) A
Barley (Hordeum vulgare) A, N
Bird'sfoot Trefoil (Lotus spp.)
Sesbania (Sesbania macrocarpa) A
Strawberry Clover (Trifolium fragiferum)
P

ADDITIONAL SPECIES

Sickle-Milkvetch (Astragalus falcatus) P
Genge (Astragalus sinicus) A or B
Borage (Borago officinalis) A, N
Mountain Bromegrass (Bromus marginatus)
P, N
Blando Brome (Bromus mollis) A, N
Pigeon Pea (Cajanus cajan) P
Garbonzo (Cicer arientinum) A
Crown Vetch (Coronilla varia) P
Silverleaf (Desmodium uncinatum) P
Horse Gram (Dolichos bifloris)
Flatpod Peavine (Lathyrus cicera) A
Ochrus (Lathyrus ochrus) A
Flat Pea (Lathyrus sylvestris var.
wagneri) P
Wimmera Ryegrass (Lolium rigidum) A, N
Greater Bird'sfoot Trefoil (Lotus major)
Succulent Lupine (Lupinus affinis) A
Small White Lupine (Lupinus angustifolius
var. diploleuca)
Cruickshank's Lupine (Lupinus cruick-
shankii) A

ADDITIONAL SPECIES

Hairy Lupine (Lupinus hirsutus) A
Green Gram (Phaseolus mungo)
Black Gram (Phaseolus mungo var. radiatus)
Giant Spurry (Spergula maxima)
Hungarian Clover (Trifolium pannonicum) P
Bitter Vetch (Vicia ervilia)

AVERAGE JANUARY
TEMPERATURE

C°		F°
20	2	68
10	3	50
0	4	32
-10	5	14
	6	

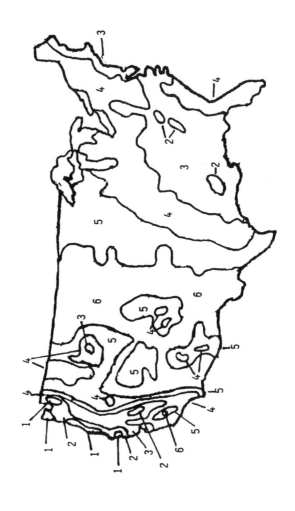

RAINFALL
November to April

cm.		in.
100	1	40
75	2	30
50	3	20
25	4	10
12½	5	5
	6	

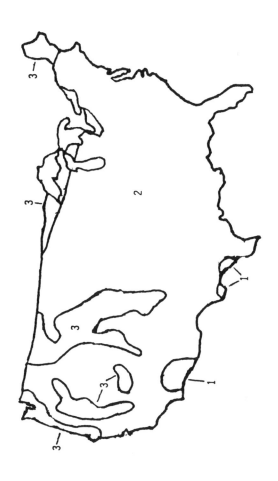

AVERAGE JULY
TEMPERATURE

C°			F°
30	1		86
20	2		68
	3		

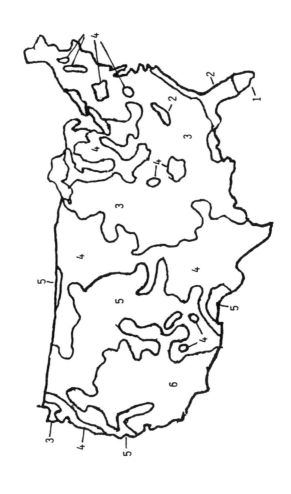

RAINFALL
May to October

cm.		in.
100	1	40
75	2	30
50	3	20
25	4	10
12½	5	5
	6	

Species Descriptions

<u>Agropyron</u> <u>repens</u> CRESTED WHEATGRASS

HABIT: Vigorous, long lived, perennial bunchgrass with an extensive root system.

USES: Forage. Crested Wheatgrass has been valuable for revegetating abandoned croplands, depleted, brushy rangeland and burned over land.

RANGE: A cool season grass. It is especially adapted to the Northern Great Plains and westward to the Sierra Nevada mountains. It is most successful in areas with 9.2 to 15.2 inches of annual rainfall. In the South, 12 to 15.2 inches of annual rainfall are required. The best altitudes for growth in the more southern areas are 4,900 to 9,200 feet.

SOIL: Crested Wheatgrass does well on most productive soils, from light, sandy loams to heavy clays. It has a low tolerance to alkaline soils and will not persist under prolonged flooding. This grass is drought resistant.

SEEDING: Sow at 6 to 12 lb.s per acre.

/h6, p6/

<u>Agrostis</u> <u>alba</u> REDTOP

HABIT: A long lived, perennial grass. It will probably thrive under a wider range of soil and climatic conditions than any other cultivated grass.

USES: Forage. To many, it is the third or fourth most important perennial grass in America, being exceeded by Timothy, Kentucky Bluegrass and perhaps Bermudagrass. Redtop is not as valuable for hay and pasture so it is cultivated only when these other grasses are not suc-

cessful. For example, it is grown where the soil is too
poor, too moist and too acid to grow Timothy.

RANGE: Redtop is found throughout the United States,
except in the drier regions and in the extreme South.
It is primarily adapted to the cooler and more humid
regions of the Northeastern U.S.. Redtop's resistance
to cold is equal to Timothy and it withstands summer
heat much better.

SOIL: Redtop thrives best on moist or wet soils. It
is one of the best wetland grasses and grows naturally
in lowlands, even when composed of peaty muck. Provided with ample moisture, Redtop does not show a marked
preference for soil type. It does best in clay loams
and loams. Redtop will grow on soils very poor in lime,
or those that are very acid. It is particularly useful
in improving impoverished clay soils. This plant is
somewhat drought reisitant. On poor uplands, even if
somewhat sandy, it will thrive better than most grasses.
Redtop is not well adapted to shade.

SEEDING: Annual Lespedezas, Ladino Clover or Alsike
Clover are usually mixed with Redtop. A recommended
seeding rate is 5 to 10 lb.s per acre.
/a5, a7, h5, h6, p4, p6, w4/

Alysicarpus vaginalis ALYCECLOVER

HABIT: A low, spreading, summer annual legume.

USES: Hay, forage and green manure.

RANGE: Alyceclover grows in the extreme South, Florida
and in the cotton belt only.

SOIL: This plant grows well in sandy loams and requires good drainage. It does not tolerate wet lands.
Alyceclover is very suseptible to nematodes so it is
often grown on clay soils.

SEEDING: A recommended seeding rate is 10 to 12 lb.s
per acre. Alyceclover reseeds itself.
/h6, m6, s9, w4/

Anthyllis vulneraria KIDNEY VETCH
HABIT: A perennial legume in the wild, but it behaves
  more like a biennial when cultivated.
USES: Pasture or hay. It produces only one harvest of
  hay per season. This legume is especially valuable
  where clover and other legumes do not thrive. It is
  less suseptible to clover rot than common clovers.
RANGE: Kidney Vetch is adapted to harsh, Northern
  climates. Drought or cold affect it little.
SOIL: It is adapted to growth on sandy or calcareous
  soils. This plant is chiefly used on dry, thin soils.
  Kidney Vetch has a long taproot which enables it to
  grow where rainfall is low.
SEEDING: Kidney Vetch can be sown under Barley, Oats or
  other grasses. It is also recommended for mixtures with
  clovers. Recommended seeding rates for this legume are:
  16 to 20 lb.s per acre drilled, or 30 lb.s per acre
  broadcast. The seed loses its germination energy
  rapidly.
  /h5, h6, p4, r1/

Arachis hypogaea PEANUT, GROUNDNUT
HABIT: An annual legume. The flowers of the peanut are
  self fertilizing and are fertilized in the bud without
  opening. Being a poor competitor, this crop is unsuit-
  able for a field with a heavy weed infestation. Peanuts
  should follow a clean cultivated crop.
USES: Peanuts have not only been grown for their
  seeds, but also for green manure, hay and pasture. The
  Spanish variety may be grown for forage on any soil
  which corn can be grown, south of the 37th parallel.
RANGE: Peanuts, when grown for seed, require at least
  five frost free months with little rainfall during
  harvest. Consequently, they are grown primarily in the
  Southern states. Generally, they need hot, moist
  growing conditions. This crop will grow at low altitudes in a latitude range between 30°N and 30°S. The
  best temperature range is from 68 to 95°F.

= 2 in.
or 5 cm.

K.B.

KIDNEY VETCH
Anthyllis vulneraria

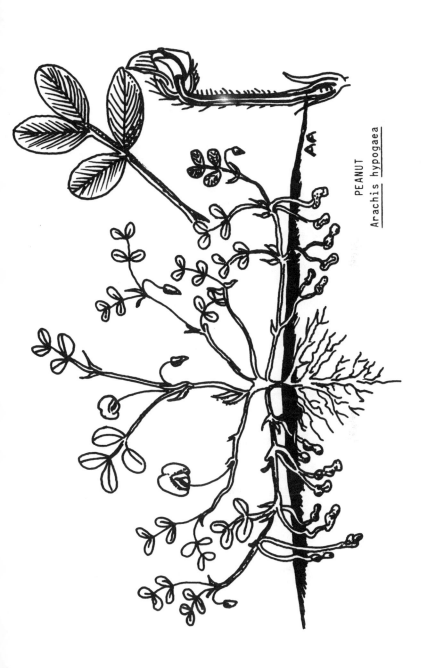

PEANUT
Arachis hypogaea

SOIL: They are adapted to many soils, but are usually
   grown in sandy and light loams. When not harvesting the
   seeds, reddish brown loams and clay loams containing an
   abundance of lime, and not too large a quantity of
   humus, are best. In poor soils, calcium or boron may
   be the nutrients limiting growth.
SEEDING: Peanuts can be interplanted with Corn or Millet,
   sowing them at the same time, or two to three weeks
   later. When grown alone, sow 40 lb.s per acre in rows
   2 to 4 feet apart. Early varieties should be sown at
   a higher density.
   /a7, h5, h6, p4, p5, s7, s8/

Astragalus cicer CICER MILKVETCH
HABIT: A perennial, rhizomatous legume which was
   introduced from Europe. It is bumblebee pollinated.
USES: Cicer Milkvetch has potential for hay and
   pasture in the Great Plains and Western states.
RANGE: This legume prefers cool, moist sites. It is
   adapted to drylands receiving more than 15.2 inches of
   annual rainfall and to wet areas at high elevations,
   with a growing season of 50 days or less. Cicer
   Milkvetch is resistant to drought, frost and pocket
   gophers.
SOIL: Sandy soils are preferred by this plant. It is
   more tolerant of alkaline or acid soils than Alfalfa.
SEEDING: Seedling emergence and growth is slower than
   that of Alfalfa or Sainfoin. A recommended seeding
   rate is 20 to 25 lb.s per acre. The seed remains
   viable for a relatively long time.
   /a9, h6/

Astragalus falcatus SICKLE-MILKVETCH, SICKLEPOD
   MILKVETCH
HABIT: Hardy, deep rooted, long lived, perennial
   legume.
USES: Forage.
RANGE: Sickle-milkvetch has been naturalized in Pullman,

Washington. It will probably withstand conditions as
severe as Alfalfa will tolerate.

SOIL: This legume is valuable on poor, dry land.

SEEDING: A recommended seeding rate is 20 to 25 lb.s
per acre. The seed remains viable for a relatively
long time.
/a9, h6, p4/

Astragalus sinicus GENGE

HABIT: A biennial or winter annual.

USES: Green manure.

RANGE: Tropics or subtropics.

SOIL: Grown only on lands that can be allowed to
become quite dry.

SEEDING: It has been sown in ripening rice.
/p5/

Avena sativa OATS

HABIT: An annual grass originating from North Africa,
the Near East and temperate Russia.

USES: Hay, pasture, green manure or cover cropping.
Land too poor to grow clover has been brought into a
clover bearing state by plowing down two crops of Oats
in the same season. Oats are good for a temporary or
rapidly growing cover crop. However, Oats may be less
desirable than Rye as a cover crop because they are
subject to winter damage. Oats are more palatable
to livestock than Rye. Due to this fact, there is
danger of stock completely destroying the value of the
cover by pasturing it to the ground.

RANGE: Oats grow best in cool, moist climates, yet
they are adapted to many climatic extremes. They
are an excellent winter cover crop in the South and in
areas where winter freezes are not severe.

SOIL: Oats are adapted to many soil types. They
are more tolerant of wet soil conditions than Barley
and require more moisture than the other small grains.
They have a low lime requirement. When fertility and

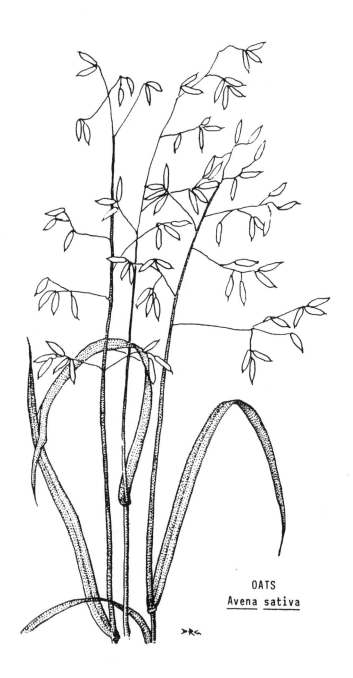

OATS
Avena sativa

drainage are moderate, this crop will tolerate a wider
pH range and a finer soil texture than Wheat or Barley.
SEEDING: Sowing can be conducted in the spring or fall
  at a recommended seeding rate of 60 to 90 lb.s per acre
  and a depth of 1 inch. A firm seedbed is believed to
  be of special value in preventing heaving of the plants
  from the soil during the winter. Oat seed retain
  their germination capacity for a relatively long time.
  /a5, a7, h6, m3, s4, w2, w3/

Borago officinalis BORAGE
HABIT: An annual, broadleaved plant which grows 1 to 2
  feet tall and blossoms all summer.
USES: Green manure and a good bee plant.
  /w3/

Brassica oleracea KALE, SCOTCH KALE
HABIT: An annual, broadleaved plant. Kale is the
  hardiest member of the cabbage family.
USES: Forage, vegetable, green manure or cover crop.
RANGE: Kale can be grown in all areas of the U.S.
  during the cool part of the year. It can be grown
  throughout the winter in the states south of New
  Jersey and well into the winter north of that line.
SOIL: This plant is adapted to a wide variety of
  soil types. It is able to grow on soils of low fertility, but has a high lime requirement.
SEEDING: In the North, Kale should be planted in the
  summer at least six weeks before the first frost. In
  the South, it can be planted into the winter. Kale is
  often interplanted with Winter Rye for soil protection
  in Northern areas. A recommended seeding rate for
  green manure is 14 lb.s per acre at a depth of $\frac{1}{2}$ inch.
  The seed remains viable for a relatively long time.
  /a5, a8, p4/

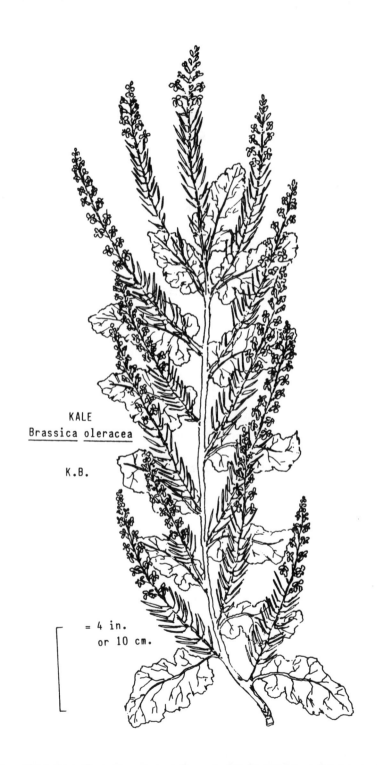

KALE
Brassica oleracea

K.B.

= 4 in.
or 10 cm.

Brassica napus RAPE, WINTER RAPE

HABIT: There are two types of Rape: the annual or summer, and the biennial or winter. Rape germinates and
grows rapidly so weeds are usually not a problem. It
matures in eight to ten weeks. This plant's long,
deep roots loosen up tough, clay soils.

USES: Forage, oil, smother crop or green manure. In
the North, Rape can serve as a Catch crop after a spring
sown crop. In the South, it can take the place of
Crimson Clover or a fall sown grain. Where Rape does
well, it makes a dense growth which smothers out many
weeds. This plant is recommended as a weed destroyer.

RANGE: Rape is best adapted to a cool growing season.
Winter Rape can be grown only where the winters are
mild, such as along the Pacific coast. It has a
climatic and soil adaptation similar to cabbage, turnips
and rutabagas, although it is less easily injured by
fall frosts.

SOIL: Loam soils that are rich, deep, moist and
contain large quantities of organic matter are best
for Rape. This plant has a low lime requirement.
However, it has been noted to grow well in calcareous
soil.

SEEDING: A recommended green manure seeding rate is
8 lb.s per acre at a depth of $\frac{1}{4}$ inch. Drilling is best
for early seeding, while broadcasting is sufficient
for later seeding. To insure a good crop, the soil must
be thoroughly plowed or disced, and then harrowed, sown
and well rolled down.
/a5, h5, p4, w3/

Brassica rapa TURNIP, COWHORN TURNIP

HABIT: A biennial, broadleaved plant. This plant is
not a good weed competitor. The Cowhorn Turnip is a
variety distinguished by its enormous, elongated root
(2 or more feet long) which constitues the bulk of
organic matter for turning under.

USES: Forage, vegetable or green manure.

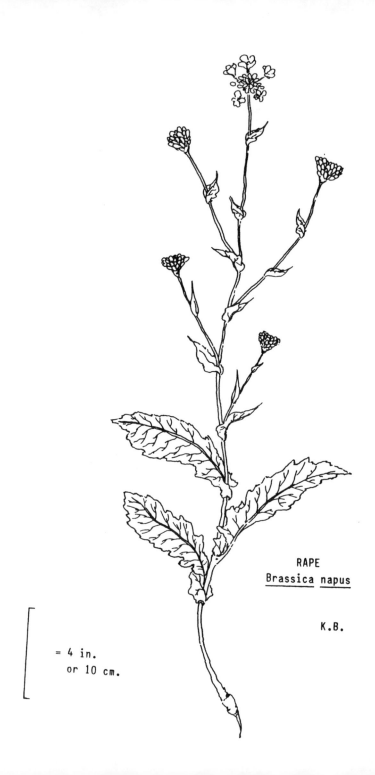

RAPE
Brassica napus

K.B.

= 4 in.
or 10 cm.

RANGE: Turnips require a cool, damp climate. They
  will not withstand drought as well as Mangel Wurzels
  (a relative of the Sugar Beet which is used for forage).
SOIL: Turnips grow well in loose, calcareous soil
  with a fine seedbed. They generally require sandier
  soils than Beets. Stiff clays are not appropriate
  because it is difficult to produce a fine seedbed in
  clay. On the other hand, light sandy or gravelly soil
  is undesirable if there is a lack of surface moisture.
  Wood ashes promote the growth of this plant more than
  the application of animal manures.
SEEDING: Turnip seed should be sown immediately after
  soil preparation because they do well in newly broken
  ground. Then the soil should be pressed down with a
  roller. Rolling helps to prevent Turnip fly damage.
  As a Catch crop, Turnips can be sown after the last
  cultivation of a Corn crop. The late summer is the best
  time to sow Cowhorn Turnips. They are not winter hardy
  so the roots are dead and decaying when turned under in
  the spring.
  /h5, p5, w3/

Bromua arvensis FIELD BROMEGRASS
HABIT: This plant is a winter annual which grows
  rapidly. It has an extensive, fibrous root system
  with a great soil holding capacity.
USES: Forage, green manure and cover crop. Field
  Bromegrass is easy to establish and makes a good winter
  cover crop. It is considered to be superior to many
  other annual cover crops.
RANGE: Field Bromegrass is adapted for growth in the
  Northeast and North Central states. It is hardier
  than Rye and is more tolerant of heat.
SOIL: It grows in any soil and has a low lime requirement.
SEEDING: Sow in either fall or spring. A recommended
  seeding rate is 15 to 30 lb.s per acre at a depth of
  $\frac{1}{2}$ inch. The seed remains viable for a relatively

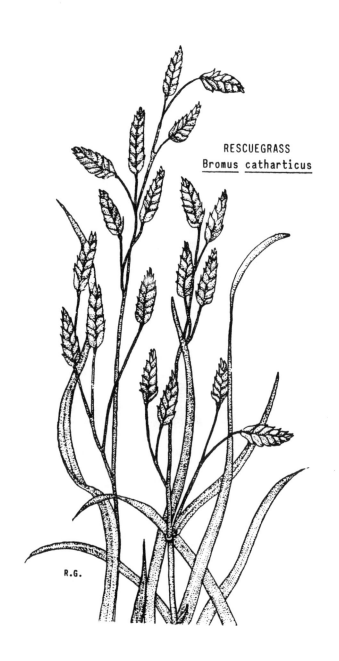

RESCUEGRASS
Bromus catharticus

R.G.

long time.
/a5, h6/

Bromus carinatus CALIFORNIA BROMEGRASS, CUCAMONGA BROME
HABIT: A vigorous, short lived, annual grass.
USES: Winter cover crop. This species could have
  extensive use for revegetation.
RANGE: It is adapted for the Pacific coast.
SOIL: California Bromegrass does well on sandy, low
  fertility soils.
SEEDING: It reseeds itself.
  /b2, w4/

Bromus catharticus RESCUEGRASS, PRAIRIE BROMEGRASS
HABIT: Rescuegrass is a cool season, short lived,
  perennial bunchgrass which behaves as a winter annual.
USES: Hay, pasture and green manure.
RANGE: Rescuegrass is best adapted to humid regions
  with mild winters. This grass has become naturalized
  in the humid sections of the Pacific coast. It is
  also adapted for growth in all of the Southern states.
SOIL: It grows in any soil type and has a low lime
  requirement.
SEEDING: A recommended seeding rate is 15 to 35 lb.s
  per acre at a depth of 3/4 inches. The seed remains
  viable for a relatively long time.
  /a5, h6, p6, w4/

Bromus inermis SMOOTH BROMEGRASS, RUSSIAN BROMEGRASS,
  AUSTRIAN BROME, HUNGARIAN BROME
HABIT: A deep rooted, sod forming, perennial grass.
USES: Pasture, hay, green manure and cover crop. Its
  fine topgrowth decomposes quickly when turned under.
RANGE: Smooth Bromegrass is a good winter cover crop
  in the North. It is adapted to cool, moist, temperate
  climates. By going dormant, this plant can survive
  periods of drought and extremes in temperature. It
  ranks between Timothy and Wheatgrass in drought

resistance.

SOIL: Smooth Bromegrass will grow in soils too dry
for Timothy or Orchardgrass, but it does not grow
nearly as well, relatively, on dry as on moist soils.
A variety of soil types are suited for its growth,
including sandy loams. Best growth is attained on
deep, fertile soils of well drained silt loam or
clay loam. It is well adapted to soils of the Western
and Northwest prairies.

SEEDING: A recommended seeding rate is 10 to 30 lb.s
per acre at a depth of $\frac{1}{2}$ inch.
/a7, a8, h6, s5/

Bromus marginatus MOUNTAIN BROMEGRASS, BROMAR MOUNTAIN
BROMEGRASS

HABIT: A short lived, deep rooted, perennial grass.

USES: Pasture and green manure. Mountain Bromegrass
mixes well with Sweet Clover for green manure.

RANGE: This species thrives in areas with cool, dry
summers and rainy winters. It is noted as being particularly well adapted to the intermountain region of
the Pacific Northwest, except where the annual rainfall is less than 16 inches.

SEEDING: A recommended seeding rate is 10 to 20 lb.s
per acre.
/h6, w4/

Bromus mollis BLANDO BROME, SOFT CHESS

HABIT: An annual grass which has a dense, fibrous
root system.

USES: Winter cover crop.

RANGE: It has been grown in the North coast area of
California.

SEEDING: It is a good reseeder.
/b2/

Bromus tectorum CHESS, CHEAT GRASS

HABIT: An annual grass.

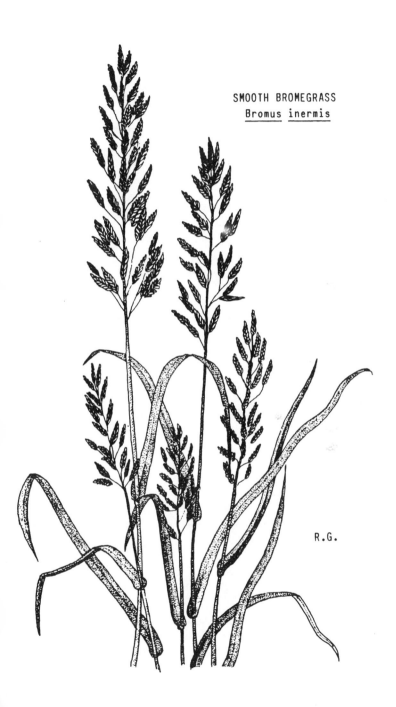

SMOOTH BROMEGRASS
Bromus inermis

R.G.

USES: Green manure.

RANGE: Winter growing in the Southwestern states.

SOIL: Chess grows best in loam soils. It has a low lime requirement.

SEEDING: A recommended seeding rate is 40 lb.s per acre at a depth of 3/4 inches.
/a5/

Cajanus cajan (C. indicus) PIGEON PEA

HABIT: A perennial legume. Pigeon Pea is a very deep rooted and drought resistant plant.

USES: Green manure, grain and pasture. In cacao and tea, Pigeon Pea has been used as a nurse crop. It can also be intercropped with plants such as Finger Millet or Corn. Pigeon Pea matures in five months. After harvest, the plants are cut back to 10 inches. The regrowth may be browsed, but if the plants are too mature, they are harmful to cattle, causing irritation of the rumen. The pods of Pigeon Pea ripen irregularly and shatter when dry.

RANGE: This legume is grown at low altitudes in the tropics and subtropics – a latitude range between 30°N and 30°S. The best temperature range is from 68 to 104°F. Pigeon Pea is suseptible to damage from frost or waterlogging.

SOIL: Pigeon Pea survives even on poor soil.

SEEDING: A recommended planting space is: 20 inches square for green manure, in rows 3 to 4 feet apart for grain or at 6.6 feet for interplanting on alternate ridges. Sow 8 to 10 lb.s per acre at a depth of 2.4 to 4 inches.
/h6, s3, s7, s8/

Canavalia ensiformis JACK BEAN

HABIT: A bushy, annual legume. It has a very deep, penetrating, root system. Jack Bean grows slowly at first. Once it is established, it is relatively fast growing, producing a crop in 3 to 4 months.

USES: Green manure, cover crop, forage and vegetable.
  After a long cooking, the green pods or ripe seeds are
  detoxified and can be eaten as a vegetable. The dried
  vegetation is too coarse and woody to make good hay.
RANGE: Jack Bean is a New World plant adapted to the
  same conditions as cotton. It grows successfully where
  the average annual temperatures range from 57 to 81 ^{o}F,
  from warmer parts of the temperate zone to hot, tropical
  rainforest areas. Early fall frosts damage the foliage,
  but not the beans. The Jack Bean grows well with annual
  rainfall as high as 165 inches and as low as 28 inches.
  They can be grown at elevations up to 5,900 feet.
SOIL: Jack Beans grow in a wide range of soil types,
  including poor soil. They grow well in acid soils
  (pH 4.3 - 6.8)and withstand waterlogged and saline
  soils. Once its deep root system is established, it
  can survive dry conditions. Full sun is required for
  its best growth, but they also grow well in shade.
SEEDING: A recommended spacing is 3 feet between rows,
  6 to 12 inches between plants. Sow at a rate of one
  bushel per acre.
  /h1, n1, p4, s7, s9/

Canavalia gladiata and C. obtusifolia SWORD BEAN
HABIT: Twining or climbing, annual legumes. They take
  6 to 10 months to mature.
USES: C. gladiata's seeds are toxic, yet the green
  pods are not and they are used as a vegetable. C. obtusifolia can be eaten by cattle.
RANGE: Sword Beans are Old World plants which are
  widely cultivated in the humid tropics of Africa and
  Asia. They grow in regions where the average annual
  temperature is between 57 and 81 ^{o}F and annual rainfall
  ranges from 28 to 165 inches. Early frosts damage the
  foliage, but not the beans. Sword Beans will grow at
  elevations up to 5,900 feet.
SOIL: Sword Beans grow on a wide range of soils, including poor and acid soils (pH 4.3 to 6.8). These

GARBONZO
Cicer arientinum

species are drought resistant. They are more resistant to waterlogging and saline soil than many other
genera. Shade is also tolerated.

SEEDING: A recommended plant spacing is 12 to 16 inches
square. Sow $4\frac{1}{2}$ to $5\frac{1}{2}$ lb.s per acre at a depth of 3/4
inches.
/n1, s7/

<u>Cicer arientinum</u> GARBONZO, CHICK PEA, GRAM PEA

HABIT: A small, quick growing, annual legume.

USES: Hay and grain. From a nutritional standpoint,
Chick Pea is one of the most valuable legumes.

RANGE: It is adapted to the tropics, but grows well
along the coast of California. At low altitudes, it
grows from 15^0 to 40^0 latitude. At high altitudes, it
grows from 0^0 to 15^0 latitude. This species can tolerate a wide range of temperatures - from 50^0 to 86^0F.
For best growth, it requires a cool growing season. It
does not withstand humidity well and prefers a rather
dry atmosphere.

SOIL: Under irrigation, Chick Pea grows successfully
during the dry season. This crop needs well drained
soil.

SEEDING: Sow in rows 3 to 4 feet apart at a rate of 20
to 30 lb.s per acre.
/h1, h6, p4, s3, s7, s8/

<u>Coronilla varia</u> CROWN VETCH

HABIT: A perennial legume with creeping stems and a
deep penetrating taproot.

USES: Temporary grazing, erosion control and green
manure.

RANGE: In the U.S., it is well adapted to the area
north of 35^0 latitude.

SOIL: This species is best adapted to fertile, well
drained soils of pH 6 or above. Once it is established,
it tolerates some degree of soil infertility and acidity.

CROWN VETCH
Coronilla varia

SEEDING: Crown Vetch can be sod seeded to improve permanent, Bluegrass pastures. It is slower in germination and seedling growth than Alfalfa and Red Clover.
   The lack of a firm seedbed and competition from
   associated grasses and weeds are the biggest problems
   in its establishment. Sow at a rate of 15 to 20 lb.s
   per acre. The seed has a medium relative longevity.
   /a9, h6, p6/

Crotolaria juncea SUNN HEMP
HABIT: An annual legume.
USES: Green manure. In an experiment conducted in
   India, Sunn Hemp yielded more nitrogen than five
   other legumes, including Cow Pea, which was the second
   highest. At eight weeks of growth, these two legumes
   yielded the same quantity of green cuttings. Nevertheless, Cow Pea may be preferable as a green manure
   because it is more leafy and less fibrous.
RANGE: Tropics and subtropics. Sunn Hemp has given
   good results in U.S. experimental plantings.
SOIL: Sunn Hemp is adapted to lighter, better
   drained lands. It does not grow as well as Sesbania
   on low, wet lands.
SEEDING: Sow at a rate of 35 to 40 lb.s per acre.
   /c1, h6, p5/

Crotolaria spectabilis (C. striata) CROTOLARIA, SHOWY
   CROTOLARIA
HABIT: A summer, annual legume. However, under almost
   frostless conditions it will continue to grow for more
   than one year. This species starts growing late and
   makes its most rapid growth in July and August.
USES: Green manure. It yields heavily.
RANGE: Tropics and subtropics. A warm season is
   necessary for its best development. Crotolaria
   requires moderate humidity. This plant has been
   reported to withstand frost injury at a minimum of
   28^{6}F.

SUNN HEMP
Crotolaria juncea

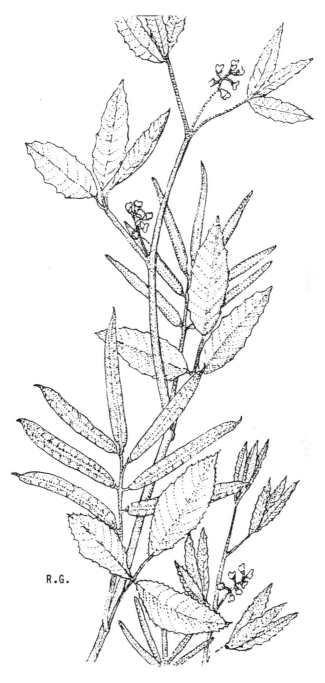

GUAR
Cyamopsis tetragonoloba

SOIL: Crotolaria is adapted to soils of low fertility.
 Requiring good drainage, it grows best in sandy or
 sandy loam soils. The lime requirement for Crotolaria
 is low. It does well on acid soils and is noted for its
 resistance to root knot nematodes.

SEEDING: Crotolaria can be sown into Oats. It will make
 most of its growth after the Oat harvest. The Oats
 serve as a nurse crop. A recommended seeding rate is
 15 to 30 lb.s per acre at a depth of 3/4 inches. The
 seed should be scarified, but if not, a larger quantity
 of seed should be sown. The seed remains viable for a
 relatively long time.
 /a5, h6, m2, m6, p2, w4/

Cyamopsis tetragonoloba GUAR

HABIT: A summer, annual legume.

USES: Green manure.

RANGE: Guar is generally adapted to the same hot and
 moist conditions as the Cow Pea. It is drought resistant.

SOIL: Guar will grow in any soil, including those of
 low fertility. It has a low lime requirement. This
 plant thrives in warm soil.

SEEDING: Avoid planting Guar too early in the spring. A
 recommended seeding rate is 30 to 40 lb.s per acre at a
 depth of 1 inch.
 /a5, h6, p4/

Cynodon dactylon BERMUDAGRASS

HABIT: A sod forming, perennial grass.

USES: Pasture, erosion control, hay and silage.
 Bermudagrass is one of the most important pasture
 grasses in the Southern states. It has a marked
 ability to withstand close grazing. This plant is an
 excellent soil binder on sandy soil or slopes. The
 rootstocks are readily eaten by hogs.

RANGE: Bermudagrass originated from India or the
 Meditteranean region. It grows in the Southern states

and in California, and is best adapted to the same
general area as cotton. Warm to hot temperatures are
best for its growth. The tops are easily killed by
frost. It is a drought resistant plant.

SOIL: This species will grow on any moderately, well
drained soil if it has an adequate supply of moisture
and plant nutrients. It grows best in clay or silt
soils, well supplied with lime. Bermudagrass will grow
better in acid, poor or dry soils than Kentucky Bluegrass. This plant is resistant to salinity and alkalinity. The coastal Bermudagrass variety will tolerate
flooding for long periods of time. It will not tolerate
shade.

SEEDING: When cultivated, Bermudagrass spreads rapidly.
It often becomes a nasty weed, particularly in row
crops. In humid regions, it produces seed sparingly.
Where it does not produce seed, it can be controlled by
first plowing it under, then growing a thickly sown and
strong growing crop such as Sorghum, Millet, Oats,
Cow Peas or Velvet Beans.

       Bermudagrass is sown by planting pieces of
rootstocks or stolons. Few plants will grow with
Bermudagrass during the summer, although Lespedeza
will hold its own. In the winter, Hairy Vetch, Bur
Clover or Italian Ryegrass can be sown into a Bermudagrass sod. As a pasture, if Bermudagrass is not
renovated through cultivation, it becomes sod bound
and loses its vigor in a few years.

/a7, h5, h6, p3, p4, p6, w4/

Dactylis glomerata ORCHARDGRASS

HABIT: A long lived, perennial bunchgrass with an
extensive root system. It begins growth in the spring
much earlier than most grasses, yet spring frosts are
injurious.

USES: Hay and orchard cover crop. When grown for hay,
it can be cut earlier than Timothy.

RANGE: Orchardgrass is grown to some extent in every

ORCHARDGRASS
Dactylis glomerata

R.G.

state. It is best adapted to the area south of which
Timothy grows best. Orchardgrass is more winter hardy
than Bermudagrass, but it is less winter hardy than
Bromegrass or Timothy. However, it will thrive under
higher temperatures and humidities than these two
grasses. The optimum temperature for Orchardgrass is
70^0F.

SOIL: This species will grow in all types of soils,
  but does not succeed well in sands or muck. It is
  best adapted to clay, clay loam or loam soils with a
  pH between 6 to 6.5. Orchardgrass needs a generous
  supply of moisture, more than Timothy, but is more
  drought resistant than Timothy. It grows well when
  irrigated. Orchardgrass is best known for its ability
  to grow well in shady places, hence its name. This
  ability is partly due to its great leafiness and partly
  to its early growth before the trees leaf out.
  Orchardgrass succeeds as well under evergreen trees as
  broadleaf trees.

SEEDING: Ladino Clover grows well with Orchardgrass. It
  does not grow well when planted alone. A recommended
  seeding rate is 6 to 15 lb.s per acre.
  /a7, b2, h5, h6, p4, p6, s5, w4/

Dalea alopecuroides DALEA, FOXTAIL, WOOD'S CLOVER
HABIT: A summer annual legume. It grows to $2\frac{1}{2}$ to
  3 feet tall.
USES: Green manure.
RANGE: Dalea is a native of the U.S.. It has been
  used successfully as a green manure for wheat in
  Western Iowa.
SOIL: Dalea grows on a wide range of soil types,
  both heavy and sandy. On strongly acid or sandy soil,
  it does well. On more fertile soils, it does not grow
  well. This legume may be useful on soils where Sweet
  Clover does not thrive.
SEEDING: Sow 10 to 15 lb.s per acre.
  /h6, p5/

Desmodium purpureum (D. tortosum, Meibomia purpurea)
  BEGGARWEED, FLORIDA BEGGARWEED, TICKCLOVER
HABIT: A summer annual legume.
USES: Green manure.
RANGE: Beggarweed is a native of the West Indies. It
  is adapted for growth in the subtropical portions of
  Florida and the Gulf states.
SOIL: Rich, moist, sandy loam soil is best for this
  legume. Nevertheless, it is not exacting in its requirements and is adapted to soils of low fertility.
  Beggarweed has a low lime requirement. Moderate rainfall is essential for its good growth.
SEEDING: Sow in the spring or early summer, after the
  danger of frost is past. A recommended seeding rate is
  8 to 20 lb.s per acre of scarified seed or 30 to 40 lb.s
  of unhulled seed. This green manure can be sown in corn
  after the last cultivation. Beggarweed will volunteer
  from year to year if the seed is allowed to mature.
  Strips of Beggarweed can be left to produce seed.
  /a5, h5, h6, m2, m6, p4, p5/

Desmodium uncinatum SILVERLEAF
HABIT: A summer growing, perennial legume.
USES: Pasture. It may be grown with Bermudagrass.
RANGE: South of central Georgia.
SOIL: Well drained.
  /h6/

Dolichos bifloris HORSE GRAM
  A leguminous, green manure crop grown in India.
  /c1/

Dolichos lablab HYACINTH BEAN, LABLAB, DOLICHOS BEAN
  AVARE
HABIT: An annual to semi perennial legume with a very
  deep root system. Hyacinth Bean retains its foliage
  into the winter longer than Cow Pea or Velvet Bean.

BEGGARWEED
Desmodium purpureum

HYACINTH BEAN
Dolichos lablab

USES: Pasture, hay, grain, vegetable, cover crop,
  smother crop and green manure. Hyacinth Bean will
  produce more hay than Cow Pea. Many cultivars can be
  grazed or cut for hay, then allowed to produce seed.
  However, you should note that this species produces a
  poor yield of seed. Nonetheless, the young pods can
  be used as a vegetable. When used for grain, the seed
  requires thorough cooking. Many cultivars are capable
  of preventing the seed of other plant species from
  germinating.

RANGE: Hyacinth Bean is adapted to practically the same
  conditions as the Cow Pea, although it has a higher
  tolerance to drought than the Cow Pea. A cool and
  moist climate can be tolerated by this plant. It is
  suseptible to frost damage. The best temperature range
  for growth of this crop is from 68 to 95°F. Arid,
  semiarid and humid regions with an annual rainfall
  ranging from 8 to 98 inches are suitable for Hyacinth
  Beans. This species has been reported to grow up to
  an elevation of 6,890 feet. At low altitudes, it will
  grow at latitudes between 0° and 30°.

SOIL: The soil requirements of Hyacinth Bean are similar to the Peanut and Cow Pea. They will grow in many
  types of soil, including some of the poorest and most
  toxic soils. For example, Hyacinth Beans will grow
  in both extremely acid or alkaline soils - a pH range
  from 4.4 to 7.8. They also grow in aluminous soils.
  Once established, it can sustain growth on light,
  sandy soil. Hyacinth Bean grows well with irrigation.

SEEDING: A recommended spacing for green manure is
  10 inches between rows and 4 inches between plants; or,
  for grain, 32 to 40 inches and 12 inches. When spaced
  for grain, a recommended seeding rate is 20 to 25 lb.s
  per acre. Good soil moisture is needed to get the
  crop established. Hyacinth Beans will reseed themselves, especially when turned under.
  /c1, h1, h6, n1, p4, s3, s7, s8/

Fagopyrum esculentum BUCKWHEAT

HABIT: A rapid growing, summer annual broadleaf.

USES: Bee plant, cover crop, smother crop, grain and
green manure. Two or three crops of Buckwheat can be
turned under in one season if you sow in the late
spring or early summer, and turn it under when in full
bloom or at eight inches in height. Its vegetation
decomposes rapidly. This plant is excellent for
increasing soil organic matter content.

  Buckwheat successfully outcompetes many noxious
weeds, such as Canada Thistle (Cirsium arvense),
because it germinates and grows rapidly, thus shading
and squeezing the weeds out. It will even smother out
Quackgrass (Agropyron repens) if the Quackgrass is thoroughly tilled before the Buckwheat is sown. For smothering Nutgrass (Cyperus spp.), grow two crops of Buckwheat, followed by Italian Ryegrass, all in one season.

  For grain, Buckwheat matures in only sixty days.
A successful grain crop of Buckwheat can generally be
grown if you sow before the middle of July. Buckwheat
is a valuable catch crop.

RANGE: Buckwheat grows well in the Northern states
during the summer months. It is sensitive to cold and
is easily killed by freezing.

SOIL: Buckwheat will grow well on any soil. This
plant is particularly valuable because it will grow on
acid soils or soils of low fertility. On hard clays,
Buckwheat's root action loosens the soil so it can be
worked more easily. Buckwheat has a low lime requirement.

SEEDING: Buckwheat should be sown only after the ground
is warm. The seeding rate varies with soil fertility.
In rich soil, sow 30 lb.s per acre, while in poor soil,
sow 50 to 60 lb.s per acre. Buckwheat is a spreading
plant. When given more room to grow, its grain yield
is higher. Therefore, when growing for grain, sow
your seed thinly.

/a2, a5, a8, m2, m6, w2, w3/

Festuca arundinacea TALL FESCUE
HABIT: An aggresive, long lived, perennial bunchgrass
  which grows during the cool season. Its heavy, fibrous
  root system penetrates the soil as much as 5 feet.
USES: Forage. Tall Fescue is more widely adaptable
  than any other cultivated forage. This species is an
  excellent soil improver, especially on heavy lands,
  because its roots are able to open up dense subsoils.
RANGE: Tall Fescue is the only cool season grass that
  will persist year after year through the hot summers
  and cool winters of California and the Southwest.
  Growth is best in the transition zone which separates
  the northern and southern regions of the U.S.. This
  grass is drought resistant, yet it will also tolerate
  high rainfall, as much as 50 to 60 inches per year.
  The Alta variety is winter hardy in Oregon. It is
  adapted to 15 inches or more rain per year and altitudes up to 5,000 feet.
SOIL: Tall Fescue will grow in any soil, except those
  that are extremely light. Best growth occurs on fertile, moist, rather heavy soil with plenty of humus.
  Poor drainage is tolerated by Tall Fescue. It will
  survive in standing water for long periods during the
  winter. This plant thrives in acid soils. It will
  tolerate moderate alkalinity and salinity.
SEEDING: Sow at a rate of 10 to 25 lb.s per acre.
  /a7, h6, p6, w4/

Glycine max SOYBEAN
HABIT: A summer annual legume which grows best in
  midsummer. Its root system is not as extensive as
  Cow Pea's and much less than Clovers'. Soybeans are
  easily crowded out by weeds when grown alone without
  cultivation.
USES: Grain, hay, silage, forage and green manure.
  Soybeans are excellent nitrogen fixers. For grain,
  Soybeans are often rotated with Corn or grown as a
  Catch crop following Wheat or Oats. Please note that

SOYBEAN
Glycine max

the seeds are poisonous when raw or improperly process-
ed. When grown for green manure, the late maturing
varieties give the largest yields.

RANGE: The Soybean is an ancient cultigen of China
and Japan. It has almost the same hot, moist, climatic
adaptations as Cotton and Corn, although it is not
adapted as far south as Corn. It grows best south of
the Kidney Bean range and north of the Cow Pea range.
Best growth occurs at temperatures between 50 to 86°F.
Soybeans withstand frost, both when young and old.
They are more drought resistant than Cow Peas and
endure dry weather better than Kidney Beans. However,
Soybeans are sensitive to a lack of water during
flowering or early pod setting.

SOIL: Soybeans will grow on any soil with good
drainage and plenty of moisture. On poor soil, they
will make satisfactory growth when inoculated, but not
as good as Cow Peas. Soybeans will grow on soils too
sandy and poor for Kidney Beans or soils too acid for
Clovers or Alfalfa. Unlike Alfalfa, Soybeans do not
require a high lime content, but a medium content.
This species does best on sandy loam, loam or clay
soils. The Mammoth variety grows well on sandy or
silty soils.

SEEDING: When the soil is thoroughly warmed, Soybeans
can be sown. A recommended seeding rate for green
manure is 45 to 60 lb.s per acre, depending on the
seed size. The best seeding depth is 1 to $1\frac{1}{2}$ inches.
If the soil dries out quickly, sow at the deeper
level. Soybean seed remains viable for a relatively
short time.
/a5, a9, h1, h5, h6, L2, m2, m6, m10, p4, p5, r1, s7,
s8, w2, w4/

Hordeum vulgare BARLEY
HABIT: An annual cereal grass.
USES: Grain, cover crop and green manure.
RANGE: Spring Barley can be planted farther north than

BARLEY
Hordeum vulgare

any other cereal. In areas with mild winter climates,
Winter Barley can be grown. For example, in the North
coast region of California, Barley can be fall sown for
cover, producing plenty of organic matter with fast
winter growth. This cereal is well adapted to high
altitudes with cold short seasons. It grows better in
dry, cool climates than hot, moist ones. Barley can
tolerate moderate droughts. It is more drought
resistant than Oats, therefore superior in semiarid
regions.

SOIL: Barley is best adapted to loam soils with good
drainage. It has a low lime requirement, but does not
grow well on acid soils. Barley can tolerate more
salinity than any other cereal. It also tolerates
alkaline soils.

SEEDING: A recommended seeding rate for green manure is
60 to 90 lb.s per acre at a depth of 3/4 inches.
Barley seed remains viable for a relatively long time.
/a5, a7, b2, h6, m5, s4, w2, w4/

Helianthus annuus   SUNFLOWER
HABIT: A summer annual broadleaf with thick stalks.
USES: Grain and green manure.
RANGE: All areas of the U.S..
SOIL: Sunflowers will grow on any soil, except those
which are acid. It has a low lime requirement.
SEEDING: A recommended seeding rate for green manure is
20 lb.s per acre at a depth of 3/4 inches.
/a5/

Indigofera hirsuta HAIRY INDIGO
HABIT: A summer annual legume.
USES: Green manure. Hairy Indigo is resistant to
root knot nematodes. Growing this legume may help to
reduce the number of root knot nematodes in infested
fields.
RANGE: Hairy Indigo grows best from Texas, along the
Gulf coast states, to Florida and the southern half of

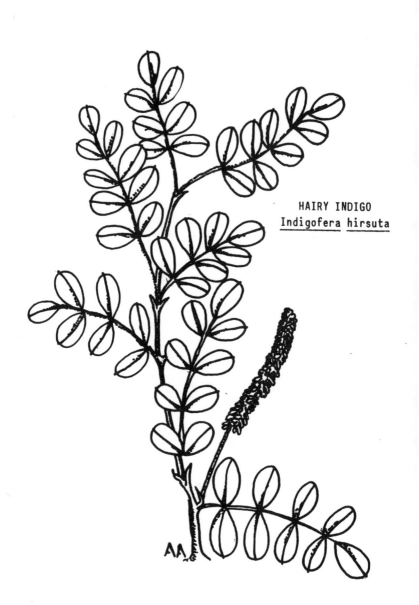

HAIRY INDIGO
Indigofera hirsuta

Georgia.

SOIL: Growth is best on sandy loams. It will also
  grow fairly well on sandy soil. This legume has a
  low lime requirement.

SEEDING: Sow Hairy Indigo in the spring or early summer
  at a rate of 8 to 10 lb.s per acre and a depth of $\frac{1}{2}$
  inch. The seed remains viable for a relatively short
  time.

  /a5, a8, h6, m6, w4/

Lathyrus cicera FLAT-PODDED VETCHLING and L. orchus
  OCHRUS

HABIT: Annual legumes.

USES: Forage.

RANGE: Both are native to the Mediterranean region.
  They may be useful along the Pacific coast.

SEEDING: Sow at 60 to 70 lb.s per acre. The seed has a
  medium relative longevity.

  /a9, h6, p4/

Lathyrus hirsutus ROUGH PEA, SINGLETARY PEA, CALEY PEA,
  SOUTHERN WINTER PEA

HABIT: A winter annual legume.

USES: Cover crop and green manure. The seeds are
  poisonous to livestock.

RANGE: Rough Pea is a native to Southern Europe. It
  grows well in the southern U.S..

SOIL: Rough Pea will grow on any soil, including
  those of low fertility. This species will grow on
  soils too wet for other winter legumes. It grows best
  on loams or sandy loams. Rough Pea has a low lime
  requirement.

SEEDING: Once established, this legume volunteers
  readily the following year. A recommended seeding
  rate is 50 to 60 lb.s per acre at a depth of 1 inch.
  Its seed has a medium relative longevity.

  /a5, h6, m4, p6, w4/

ROUGH PEA
Lathyrus hirsutus

R.G.

Lathyrus sativus GRASS PEA
HABIT: An annual legume.
USES: Forage.
RANGE: Cool, moist growing conditions. It thrives
  wherever the Field Pea can be grown.
SOIL: Well drained clay loam, loam and sandy loam
  soils.
SEEDING: Sow at a rate of 70 to 80 lb.s per acre. The
  seed has a medium relative longevity.
  /a9, h6, p4/

Lathyrus sylvestris var. wagneri FLAT PEA, WAGNER PEA
HABIT: A long lived, rhizomatous, perennial legume
  which resembles the Field Pea.
USES: Cover crop and erosion control. Flat Pea is
  recommended for use as a permanent orchard cover. It
  is not a good forage because the seeds are toxic to
  grazing animals.
RANGE: It is well adapted for growth in the Northern
  states.
SEEDING: A recommended seeding rate is 60 to 70 lb.s
  per acre. The seed has a medium relative longevity.
  /a9, h6, p4/

Lathyrus tingitanus TANGIER PEA
HABIT: A winter annual legume.
USES: Forage and green manure. Tangier Pea can
  choke out weeds.
RANGE: Cool, moist climate. It is adapted to the
  same growing conditions as the Field Pea. This legume
  is grown for seed in the fog belt of California
  because the pods shatter readily in sunny, arid regions.
SOIL: Clay loam, loam or sandy loam soils which are
  well drained. Tangier Pea has a medium lime requirement.
SEEDING: Sow 70-80 lb.s/acre at a depth of 1 inch in the
  spring, or in areas with mild winters, sow in the fall.
  Frost effects them less than other winter legumes. The
  seed has a medium relative longevity.
  /a5, a9, h6, k3, p4, p5/

LENTIL
Lens esculenta

Lens esculenta LENTIL
HABIT: An annual legume.
USES: Grain.
RANGE: Lentils grow in temperate and subtropical regions during the cool, moist season of the year. They are usually grown under natural rainfall. A $3\frac{1}{2}$ month growing season is required for this crop to mature. At low altitudes, Lentils grow at latitudes between 15°N and 40°N. Growth of this crop occurs at temperatures ranging from 36 to 86°F.
SOIL: They will succeed on any soil.
SEEDING: Sow at a rate of 11 to 15 lb.s per acre, when spaced in rows 3 to 4 feet apart. The seed has a medium relative longevity.
/a9, h6, s3, s7/

Lespedeza cuneata SERICEA LESPEDEZA
HABIT: A hardy, summer growing, perennial legume. It is similar in growth to Alfalfa.
USES: Forage, cover crop and green manure. Sericea Lespedeza greatly increases the yields of succeeding crops.
RANGE: When provided with adequate moisture, Sericea Lespedeza grows in all parts of the U.S., except the Northwest.
SOIL: Sericea Lespedeza will grow on soils of greater acidity and lower fertility than most crops. It has been used successfully on poor, clayey soil and on slopes and banks with soil too poor to support most other cover crops. This legume prefers loam soils and has a low lime requirement.
SEEDING: It is easy to establish on hard and badly eroded soils. A recommended seeding rate for green manure is 25 lb.s per acre at a depth of $\frac{1}{2}$ inch. Sow early in the spring. The seed has a medium relative longevity.
/a5, h6, m3, m6, p6/

COMMON LESPEDEZA
Lespedeza striata

Lespedeza stipulacea KOREAN LESPEDEZA
Lespedeza striata COMMON LESPEDEZA, JAPANESE LESPEDEZA,
  JAPAN CLOVER

HABIT: Summer annual legumes. Spring growth is slow
  until warm weather arrives. They make good growth in
  the hot weather of late summer and fall, when most
  other crops are mature. Korean Lespedeza is an
  earlier maturing species than Common Lespedeza and
  will mature farther north.

USES: Hay, pasture and green manure. Annual Lespedezas are well suited for use in rotations with winter
  cereals because of their late summer growth. However,
  they are not good weed competitors due to the short-
  ness of their growing season.

RANGE: Korean and Common Lespedeza are well adapted
  for growth in the Southeast and Gulf coast states.
  Common Lespedeza requires a longer growing season than
  Korean so is adapted farther south. Both species are
  strongly drought resistant. Neither of them will
  withstand frost.

SOIL: The Annual Lespedezas will grow in any, well
  drained soil, except in extremely sandy soil, when
  provided with sufficient moisture. For best growth,
  they prefer loams. They will also grow in eroded
  soils, acid soils, soils low in phosphorus or of
  low fertility. On soils too acid to grow Clover
  without the use of lime, Lespedezas will grow. Korean
  is less tolerant of acid soil and more tolerant of
  alkaline soil than Common. Both have a low lime
  requirement.

SEEDING: Early seeding is necessary in order to obtain
  a good stand. A recommended seeding rate is 10 to 20
  lb.s per acre of unhulled seed at a depth of $\frac{1}{2}$ inch.
  The seed remains viable for a relatively short time.
  These species are known to reseed themselves and
  volunteer the next year.
  /a5, a7, a8, h5, h6, m2, m3, m6, p4, p6, w4/

Lolium multiflorum ITALIAN RYEGRASS, ANNUAL RYEGRASS
HABIT: Italian Ryegrass is not truely an annual, but,
  under farm conditions, few of the plants live more
  than one year so it is generally considered to be one.
  It has a heavy, fibrous root system which holds the
  soil well. This plant is well known for germinating
  and growing rapidly.
USES: An important cover crop and green manure. It
  can also be used for hay and forage. Being a rapidly
  growing plant, it provides quick cover and excellent
  temporary pasture.
RANGE: Italian Ryegrass grows in all parts of the U.S.,
  but is best adapted to the same area as Crimson Clover.
  It is not as winter hardy as Timothy or Orchardgrass.
SOIL: Italian Ryegrass will grow on any soil. It
  grows best on loams or sandy loam soils. On well
  drained land, this plant is adapted to irrigation
  farming. Short periods of flooding will not damage
  a stand of Italian Ryegrass, although long periods
  will. This crop has a low lime requirement.
SEEDING: A recommended green manure seeding rate is
  25 to 35 lb.s per acre at a depth of 3/4 inches. In
  the South and on the Pacific coast, sow it in the
  fall. While in the North, sow it in the spring to
  avoid freezing. The seed remains viable for a relatively long time.
  /a5, b2, h6, p3, p4, p6, w4/

Lolium perenne PERENNIAL RYEGRASS, ENGLISH RYEGRASS
HABIT: A short lived, rapidly growing, perennial
  grass. It lives only two years on poor land.
USES: Forage.
RANGE: Perennial Ryegrass is adapted to regions with
  mild, moist winters, such as in the Southern states,
  Pacific coast states, and somewhat east of the

ITALIAN RYEGRASS
Lolium multiflorum

Cascade mountains. This plant continues to grow at
low temperatures, but does not withstand severe winter
cold. It is about equal to Orchardgrass in winter
hardiness and is more winter hardy than Italian Ryegrass. Shading by other grasses is not tolerated by
Perennial Ryegrass.

SOIL: Perennial Ryegrass will grow in any soil,
provided it is moist and well drained. It will not
endure standing water near the soil surface. Growth
is about equal to other grasses on soils of low
fertility.

SEEDING: A recommended seeding rate is 25 to 35 lb.s
per acre. When grown on poorer soils, heavier sowing
is required.
/h6, p4, w4/

Lolium rigidum WIMMERA RYEGRASS, SWISS RYEGRASS

HABIT: A winter annual.

USES: Pasture and cover crop. In the North coast
region of California, this species has been sown as
an orchard and vineyard cover crop.

RANGE: Wimmera Ryegrass grows in regions with mild,
moist winters.

SOIL: This grass grows well in wet spots. In
Australia, it was found to be valuable on poor soil.

SEEDING: It will reseed itself.
/b2, p4/

Lotus corniculatus BROADLEAF BIRD'SFOOT TREFOIL
Lotus tenuis NARROWLEAF BIRD'SFOOT TREFOIL

HABIT: Warm season, low growing, perennial legumes.
They have a well developed tap root which penetrates
deeply into the soil and becomes woody with age. These
species are honey or bumblebee pollinated. They are
almost completely self sterile.

USES: Pasture, hay and silage.

RANGE: These two Bird'sfoot Trefoil species grow in
all parts of the U.S.. However, they do not persist

BROADLEAF BIRDSFOOT TREFOIL
Lotus corniculatus

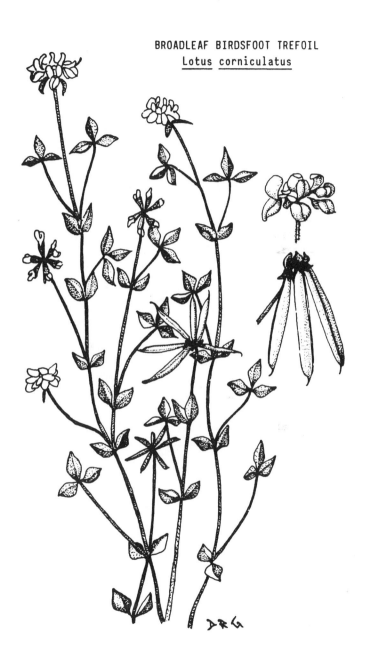

well in most parts of the Southeast. Both species are
resistant to high temperatures and drought.

SOIL: The Lotus species are characterized by growing
   in the driest situations and upon the lightest and
   most sterile soils. Broadleaf and Narrowleaf will
   thrive on almost any soil. They will grow on land
   too poor, shallow, impermiable or deficient in lime to
   support Red Clover or Alfalfa. Fertile, moist,
   moderate to well drained soils with a pH from 6.2 to
   6.5 are best for their productivity, yet these species
   are generally used on poorer, drier soils.

        Broadleaf grows better on dry soils than Narrow-
   leaf. These species respond well to phosphate or
   potash fertilization. Both species are outstandingly
   salt tolerant. They are also valuable legumes because
   they are resistant to dodder and broomrape.

SEEDING: When sown alone, a recommended seeding rate is
   10 to 18 lb.s per acre broadcast or 9 to 12 lb.s per
   acre drilled. They are often sown with grass species.
   The seed has a medium relative longevity.
   /a7, a9, h5, h6, p4, p6, r1, w4/

Lotus major GREATER BIRD'SFOOT TREFOIL
        This legume is found only where there is plenty
   of moisture. It is also fond of shade. Marshy or
   peaty land can be improved by growing Greater Bird'sfoot
   Trefoil.
   /r1/

Lotus pendunculatus (L. uliginosus) MARSH BIRD'SFOOT
   TREFOIL, BIG TREFOIL
HABIT: A perennial legume.
USES: Forage.
RANGE: In the U.S., Marsh Bird'sfoot Trefoil is known
   to grow in wet, poorly drained soils of the Northwest
   coast and also along the Atlantic and Gulf coasts. In
   the Northwest, this species has become naturalized.
SOIL: As its name implies, Marsh Bird'sfoot Trefoil

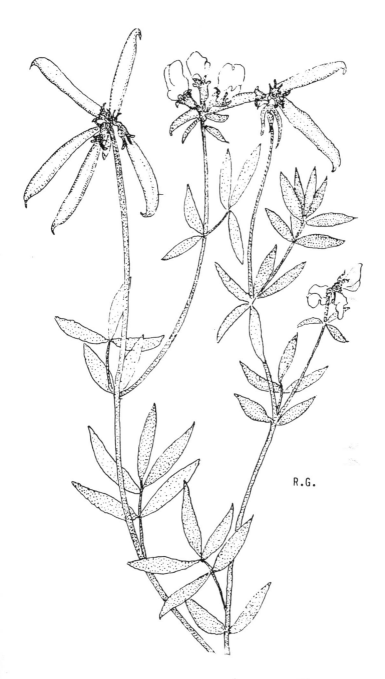

R.G.

NARROWLEAF BIRDSFOOT TREFOIL
Lotus tenuis

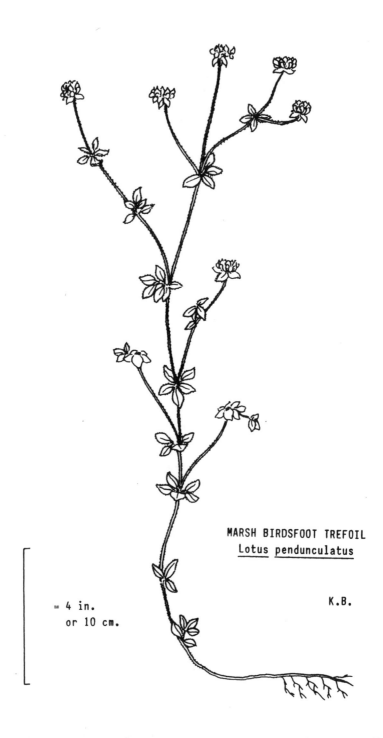

MARSH BIRDSFOOT TREFOIL
Lotus pendunculatus

K.B.

= 4 in.
or 10 cm.

grows well in wet soils, such as in marsh or moor land,
along ditches and in shady situations. This species
can withstand long periods of surface flooding after
the plants are well established. It rarely grows on
hills, except in areas with high rainfall. Soils of
pH from 4.5 to 6.0 seem to be best for its growth.
/a7, h5, h6, w4/

Lupinus affinis SUCCULENT LUPINE
        Succulent Lupine is an annual legume, indige-
nous to California. Compared to other Lupines, it is
slow in maturing and less resistant to frost, but it
does produce a heavy seed crop. This legume also
produces an abundance of root nodules, both in new and
manured soil. Succulent Lupine decomposes less readily
than Yellow Lupine.
/d1/

Lupinus albus LARGE WHITE LUPINE
HABIT: Large White Lupine is a rapidly growing, annual
  legume. Its roots penetrate the soil as far as two
  feet deep.
USES: Green manure, cover crop, forage, grain, mulch
  and smother crop. Large White Lupine is an ancient,
  green manure plant which improves the poorest soil.
  It is excellent for the renovation of worn out soils
  and the improvement of soils that are naturally barren.
  For sheep, this legume makes an excellent fodder, green
  or dry. The seeds of this plant can be eaten, after
  they have been boiled to eliminate poisons. Large
  White Lupine has also been used as a mulch around trees.
  It is sometimes grown to smother out noxious weeds.
RANGE: Large White Lupine is best adapted for growth in
  Florida, the Gulf coast, the Northeast and Northcentral
  states. This is the most winter hardy Lupine species.
  The "Hope" variety has been developed specifically as
  a winter cover crop and green manure for Arkansas.
SOIL: Large White Lupine will grow in any soil,

except those which are marly or calcareous. It is best
adapted for growth in sandy loam soils and also grows
well on land with subsoils rich in iron. This plant
is useful in opening up stiff, clay soil. It is also
one of the best green manures for transforming sandy
soil into a productive state. This species has a low
lime requirement. It grows well in acid soils.

SEEDING: Sow in the fall or early spring. A recommended
green manure seeding rate is 100 to 120 lb.s per acre
at a depth of 1 inch. The seed remains viable for a
relatively short time.

/a5, d1, h6, w3/

Lupinus angustifolius NARROW-LEAVED LUPINE

Narrow-leaved Lupine is an annual legume which
can be utilized as a forage for sheep. It is often
found as a weed in the grain fields of the Mediterranean
region, Southern Europe and Northern Africa. This plant
grows well in poor, sandy soil. It does not tolerate
soils with an excess of lime. This species has some
unfavorable characteristics: it is liable to root rot,
will not decompose readily, produces few nodules and
flowers late. Nevertheless, it can't hurt to experiment
with it.

/d1/

Lupinus angustifolius var. caeruleus SMALL BLUE LUPINE

HABIT: An annual legume. Small Blue Lupine produces
an abundance of seed. The seed pods do not break up
as readily when harvested as Yellow Lupine's.

USES: Green manure. When turned under, this plant
fails to rot thoroughly and quickly. It used to be
fed to sheep, but this practice was stopped due to the
plant's poisonous qualities.

RANGE: Small Blue Lupine is best adapted for growth in
Florida, California, and the Southeastern and Gulf
coast states.

SOIL: This Lupine is fairly tolerant of calcareous

NARROW-LEAVED LUPINE
Lupinus angustifolius

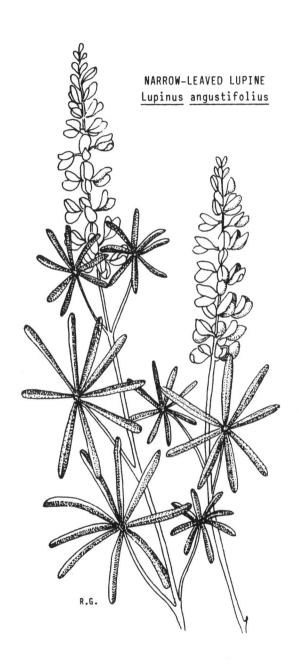

R.G.

soils in California, but is less so than <u>L</u>. <u>pilosus</u>.
In the coastal plain of the Southeast, it grows best in
neutral to slightly acid soils of at least moderate
fertility. It grows very well on stiff and clayey soil
which is well prepared by repeated plowing. On clay
soil, Small Blue Lupine grows better than Yellow
Lupine. On granitic and red soil, it grows poorly.
Small Blue Lupine is suseptible to root rot.
SEEDING: A recommended seeding rate is 70 to 100 lb.s
per acre at a depth of 1 inch. The seed remains
viable for a relatively short time, even shorter than
Yellow Lupine.
/a5, d1, h6/

<u>Lupinus</u> <u>angustifolius</u> var, diploleuca SMALL WHITE
LUPINE
        As a forage, this variety is less injurious
to livestock than Small Blue Lupine. It is equally
suseptible to root rot.
/d1/

<u>Lupinus</u> <u>cruickshanksii</u> CRUICKSHANK'S LUPINE
        Cruickshank's Lupine is an annual legume. It
is preferred by livestock over Yellow Lupine, especially
when grown with Clover. Its stem is woody.
/d1/

<u>Lupinus</u> <u>hirsutus</u> HAIRY LUPINE
        Hairy Lupine is an annual legume used for
forage. Cattle prefer its vegetation and seeds to
either Small Blue Lupine's or Yellow Lupine's. This
plant requires good soil and will not thrive on poor,
sandy soil. The seeds ripen late and the pods open
too easily.
/d1/

Lupinus luteus var. sativus YELLOW LUPINE, FRAGRANT
  LUPINE
HABIT: An annual legume with fragrant blossoms.
  Yellow Lupine's roots penetrate deeper into the soil
  than Small Blue Lupine's. This species resists frost
  and root rot.
USES: Hay and an excellent green manure. This
  species is the least winter hardy and least bitter of
  the Lupines. Animals do not care for it in a green
  state so it is made into hay. It was once extensively
  cultivated as a green manure to improve poor, sandy
  soil.
          Yellow Lupine has been grown in rotation with
  Potatoes and Winter Cereals in Hanover, Germany. At
  the last cultivation of the Potato crop (April –
  middle May), Yellow Lupine was sown between the potato
  rows. In the fall, the Potatoes were harvested and
  Lupines turned under, then Rye or Winter Wheat were
  sown. In August, the grain was harvested. Then
  Yellow Lupine was sown again, plowed under in the
  spring and Oats sown.
SOIL: Yellow Lupine grows well in sandy soil.
  Exhausted, sandy soils, once worthless, have been
  brought back into production by green manuring with
  Yellow Lupine. It will even thrive on coastal drift
  sand. This legume also grows well in sandy loam soils,
  light red soils rich in iron, and acid soils. Yellow
  Lupine requires good drainage and does not tolerate
  standing water. If the soil is suitable, Yellow Lupine
  always outcompetes the weeds.
SEEDING: A recommended seeding rate is 70 to 110 lb.s
  per acre at a depth of 1 inch. Sow either in the fall
  or the spring, depending upon the severity of winter.
  The seed remains viable for a relatively short time.
  /a5, a8, d1, h6, s2, w3/

Lupinus perennis PERENNIAL LUPINE, SUNDIAL LUPINE
HABIT: A perennial legume with creeping rootstocks

which draw surface moisture.

USES: Forage. As a pasture plant, it is less object-
ionable to animals than Yellow Lupine. It is not
suited for use as a green manure or cover crop in an
orchard or vineyard.

SOIL: Clay soil which retains surface moisture.
This species may be valuable where the surface soil is
good and the subsoil poor.
/d1/

Lupinus pilosus var. caeruleus LARGE BLUE LUPINE
Lupinus pilosus var. roseus PINK LUPINE

HABIT: Annual legumes. These species produce a large
number of root nodules, even under adverse conditions
which will cause an absence in other Lupine species.
They are not affected by frost or root rot.

USES: Green manure. When plowed under at first
flowering, they decay rapidly.

SOIL: Large Blue Lupine is the best Lupine for heavy,
calcareous soil.
/d1/

Lupinus termis EGYPTIAN LUPINE

The seeds of Egyptian Lupine can be eaten after
steeping. It has a woody stem and flowers later than
Large White Lupine. This species grows well in cultivated, sandy loam soil. It benefits from manure and a
small quantity of lime.
/d1/

Medicago arabica (M. maculata) SPOTTED BUR CLOVER,
SPOTTED MEDIC

HABIT: A winter annual legume.

USES: Forage and green manure.

RANGE: Spotted Bur Clover is adapted to regions with
mild, moist winters. The Bur Clovers cannot endure
heat. In the Southern states, Spotted Bur Clover is
more plentiful than Toothed Bur Clover. On the

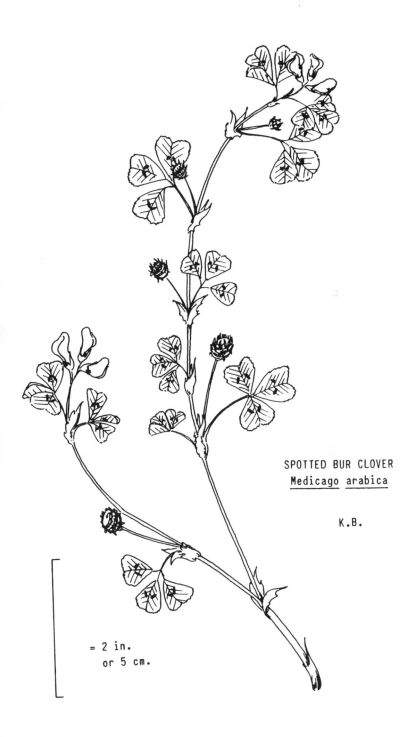

SPOTTED BUR CLOVER
Medicago arabica

K.B.

= 2 in.
or 5 cm.

Pacific coast, it is abundant only along the coast, along streams and in shady places. Spotted is more resistant to cold than Toothed, withstanding temperatures down to 15°F.

SOIL: Any soil.

SEEDING: Sow 30 lb.s per acre of hulled seed, or 100 lb.s per acre of unhulled seed at a depth of ½ inch.
/a5, h5, h6, m5, p4, p5, p6, w4/

Medicago hispida (M. polymorpha, M. denticulata)
TOOTHED BUR CLOVER, TOOTHED MEDIC, BUR CLOVER, CALIFORNIA CLOVER

HABIT: A winter annual legume. Toothed Bur Clover is adapted to drier and sunnier situations than Spotted Bur Clover.

USES: Green manure, cover crop, forage and hay. This winter growing legume can be rotated with a summer row crop. For example, Bur Clover has been grown as a winter cover crop and green manure in cotton fields. This practice increases cotton yields inexpensively. It is also commonly used as a green manure crop in the orchards of California, producing volunteer crops year after year. Toothed Bur Clover is readily eaten by sheep, although their spiney seeds get caught in the sheep wool.

RANGE: Being a native of the Mediterranean region, Toothed Bur Clover is adapted to mild, moist winters. This species will grow in all parts of the U.S., except the Northeast and Northcentral states. It grows best in California.

SOIL: This species will succeed on any type of soil. Best growth occurs in well drained, loam or clay loam soils, rich in lime. Toothed Bur Clover will grow on soil too deficient in lime to support Crimson Clover. It will succeed in somewhat acid soils, but does best in neutral or slightly alkaline soils.

     Where the soil is poorly drained and very moist, the plants mature much later than on well

TOOTHED BUR CLOVER
Medicago hispida

DRG

drained land. In low rainfall years, if the winter temperatures are not too cold, Toothed Bur Clover will produce plants.

Toothed Bur Clover does poorly on soils of low fertility. Fertilizing will help to establish stands on poor soil. Phosphate is the most essential nutrient.

SEEDING: Toothed Bur Clover readily reproduces itself with little or no reseeding. It is possible to have a good stand of Bur Clover each year, without annual seeding, by allowing a Bur Clover crop to mature once every four or five years. In that year, after turning the Bur Clover under, a crop of late Corn can be sown as the season's cash crop. This system works well with cotton.

Be sure you want this plant for many years before sowing because it does reseed itself very well. A recommended seeding rate is 20 to 30 lb.s per acre at a depth of $\frac{1}{2}$ inch. The seed remains viable for a relatively long time.

/a5, a8, b2, d1, h5, h6, L1, m3, m4, m5, m11, p4, p5, w4/

Medicago lupulina BLACK MEDIC, YELLOW TREFOIL

HABIT: An annual or semi-biennial legume.

USES: Forage and green manure. Black Medic is not very palatable to livestock, yet it is nutritious. It is never sown alone, but in mixtures with Clovers and grasses. This legume does not grow well with Red Clover. Where Clovers fail, try Black Medic. It is less suseptible to the Sclerotinia fungus than Red Clover.

RANGE: There are many varieties of Black Medic which grow under different climatic conditions. Generally, Black Medic grows best in a cool, moist climate. It has a climatic adaptation similar to Alsike Clover. Black Medic will withstand much greater cold than Crimson or Red Clover.

SOIL: Black Medic will grow on most soils. Best

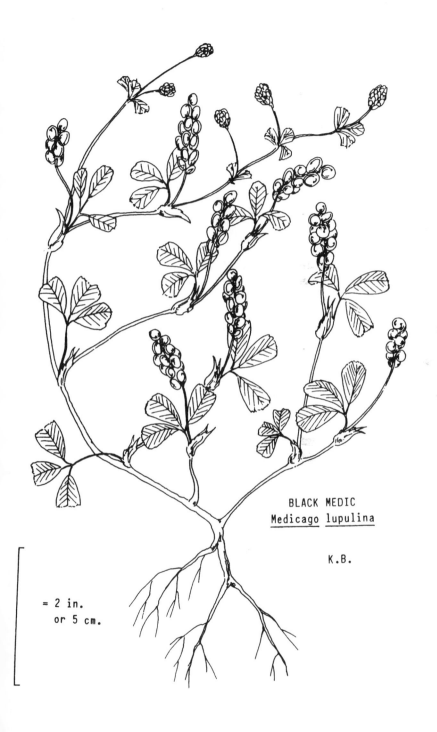

BLACK MEDIC
Medicago lupulina

K.B.

= 2 in.
or 5 cm.

growth occurs on calcareous soils, or reasonably fertile
lands, not distinctly acid, which are fairly well
supplied with moisture. On poor, mildly acid soils,
Bur Clover grows better than Black Medic. Applying
lime will help it grow on poor soil.

SEEDING: Sow in the late summer or early fall. A
recommended seeding rate is 10 to 15 lb.s per acre.
The seed remains viable for a relatively long time.
/a9, h5, h6, p4, p5, r1, w4/

Medicago orbicularis BUTTON MEDIC, BUTTON CLOVER
HABIT: A winter annual legume.
USES: Forage and green manure. It is an excellent
companion with winter growing, small grains.
SOIL: Button Medic has a wide soil adaptation. It
grows best in fertile, shallow, calcareous soil.
SEEDING: Sow at a rate of 15 to 20 lb.s per acre.
/h6, p6/

Medicago sativa COMMON ALFALFA
HABIT: A summer perennial legume with deep roots.
USES: Hay, forage, grain for sprouting, cover crop
and green manure. As a hay crop, Alfalfa can last five
years or longer. This plant is an excellent cover
crop in orchards, especially irrigated apple orchards
of the Rocky mountain and Pacific coast states. In an
orchard, leave the entire crop for mulch or cover. For
trees under four years old, make sure the trees get
enough water because Alfalfa has deep roots and will
take up much of the moisture. Alfalfa is one of the
best legumes for improving the physical condition of
the soil and adding nitrogen. In the Southwest,
Alfalfa is the best crop to precede lettuce, canta-
loupe and similar crops.
RANGE: Alfalfa is best adapted to a warm, dry, semiarid
climate, or one with moderate rainfall. This plant is
not well adapted to humid conditions. High temperatures
with accompanying moderate humidity are very injurious.

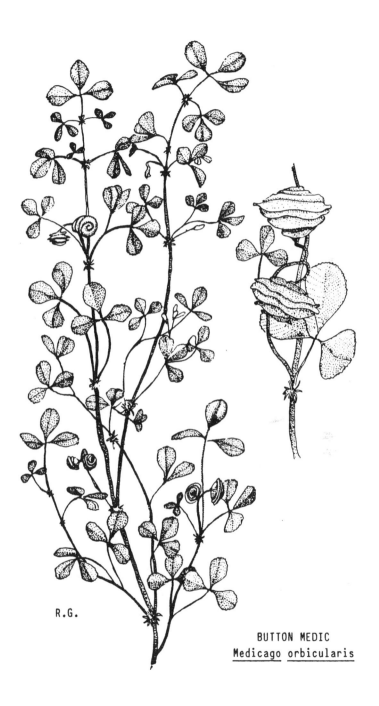

R.G.

BUTTON MEDIC
Medicago orbicularis

Alfalfa grows best in a relatively dry atmosphere with
irrigation.

For almost every region of the U.S., there is
a variety of Alfalfa. Growth of these varieties range
from below sea level in California, to an elevation of
8,000 feet in Colorado. This plant is highly drought
resistant and withstands hot weather. On the other
hand, Alfalfa does not endure cold weather. However,
it does endure frost.

SOIL: Alfalfa grows in any, deep, well drained soil.
Its roots won't grow into water logged soils, nor will
they penetrate a thick hardpan. Where the topsoil and
subsoil are loose, friable and porous, and contain a
large quantity of lime, Alfalfa will thrive.

Alfalfa tolerates alkali and salt concentrations
better than most crops. On strongly alkali soils, as
are often found in the West, it makes little or no
growth. Sweet Clover is more tolerant in this respect.

This legume is sensitive to pH, rarely growing
well at a pH below 6. The most favorable pH is between
6.5 and 7.5. Add lime to acid soils for a successful
Alfalfa crop. In addition, a good supply of potassium
and phosphorus are also necessary for its best growth.

SEEDING: A recommended seeding rate is 15 to 20 lb.s
per acre at a depth of $\frac{1}{2}$ inch. The seed remains viable
for a relatively long time. Sow Alfalfa in a firm
seedbed during the late winter or early spring. It
makes a better stand when drilled, than broadcast.
At first, the plants spend most their energy developing
roots and are easily smothered by weeds, so cultivation
is necessary.

/a7, a8, a9, h5, h6, L2, L5, m2, m6, m10, p4, p5, r1,
w4/

Melilotus alba WHITE SWEET CLOVER, BOKHARA SWEET CLOVER
Melilotus officinalis YELLOW SWEET CLOVER, OFFICIAL
   MELILOT, COMMON SWEET CLOVER
HABIT: Summer growing, biennial legumes with deep

R.G.

COMMON ALFALFA
Medicago sativa

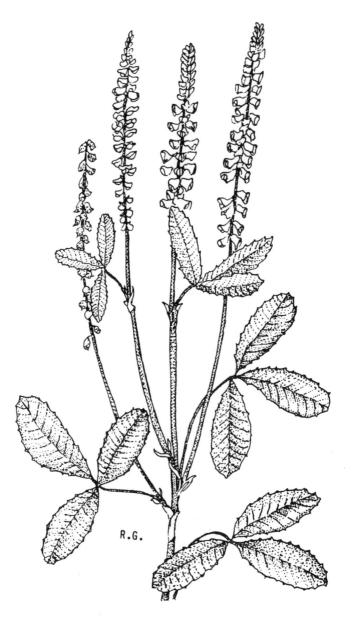

R.G.

WHITE SWEET CLOVER
Melilotus *alba*

YELLOW SWEET CLOVER
Melilotus officinalis

R.G.

roots.   There are annual varieties of <u>Melilotus</u> <u>alba</u>.
All Melilotus species contain a vanilla smelling substance, coumarin, which makes the older stems and
leaves quite bitter. Yellow Sweet Clover is an
earlier blooming, smaller plant than White Sweet
Clover.

USES: Hay, pasture and green manure. Sweet Clovers
are recommended for renovation of very poor, thin
soils. They fit well in a rotation with a Cereal
and Corn. The Sweet Clover is sown with the Cereal,
Wheat, Oats or Barley, for example. Even Flax can be
used in this manner. They act as nurse crops for the
Sweet Clover.

When turning Sweet Clover under for green
manure, always cultivate in the early spring, rather
than the fall. Sweet Clover is a biennial so it will
resume growing the second year if turned under in the
fall. Spring cultivation prevents the Sweet Clover
from continuing growth.

RANGE: Sweet Clovers will grow almost anywhere,
provided there is more than 17 inches of rain,
suitably distributed throughout the year. They grow
in southern Canada and throughout the entire U.S.,
thriving equally well in semiarid and humid regions.
Sweet Clovers often do well in areas too dry for
Alfalfa to succeed. Yellow Sweet Clover withstands
drier conditions than White.

SOIL: The Sweet Clovers will grow on a wide range of
soils, from cemented clays and gravels to poor sand.
Their long roots open the subsoil. Best growth
occurs on neutral or alkaline soil with an abundance of
lime. If the pH is below 6, lime must be applied well
in advance of seeding. Yellow grows best in loam soils,
while White prefers clay loams.

There is one characteristic which makes Sweet
Clover extremely valuable. They have a greater ability
to extract nutrients, phosphorus and potassium for
example, from insoluble minerals than most other crops.

SEEDING: Sow during the fall or spring in well prepared, firm soil. However, please note that White germinates better from fall sowing. A recommended seeding rate is 10 to 15 lb.s per acre of scarified seed or 25 lb.s per acre of unscarified seed at a depth of $\frac{1}{2}$ inch. Unscarified seed germinates better when sown in the fall. The plants grow slowly at first. Sweet Clover seed remains viable for a relatively long time.
/a5, a7, a8, a9, h5, h6, L2, m2, m3, m6, m7, p4, p5, r1, w4/

<u>Melilotus indica</u> SOUR CLOVER, BITTER MELILOT, ANNUAL YELLOW CLOVER, BITTER CLOVER, KING ISLAND MELILOT

HABIT: A winter annual legume. Sour Clover has a woody stalk and a long, deep taproot. It is a heavy seed producer. Quite often, it is a prevalent weed in Wheat. In fact, the seed of Sour Clover can be collected from screenings of infested Wheat.

USES: Green manure and cover crop. This legume has been used often as a green manure and cover crop in the orchards of southern and central California, particularly Citrus. Experiments conducted in southern California have shown that Sour Clover outyields Common Vetch and Canadian Field Peas. In addition, Sour Clover withstands more frost than these two. It is also resistant to aphids.

RANGE: Anywhere in the U.S.. This species is not winter hardy. It grows well in the Southwest.

SOIL: Almost any soil. Sour Clover grows best in loams. Its long, large taproot helps break up heavy soils. This species has a high lime requirement.

SEEDING: A recommended seeding rate is 10 to 15 lb.s per acre at a depth of $\frac{1}{2}$ inch. In Southern California, Sour Clover should be planted no later than October 15th if it is to be plowed under by March. The seed remains viable for a relatively long time.
/a5, a9, d1, h6, L2, m8, p4, r1, w4/

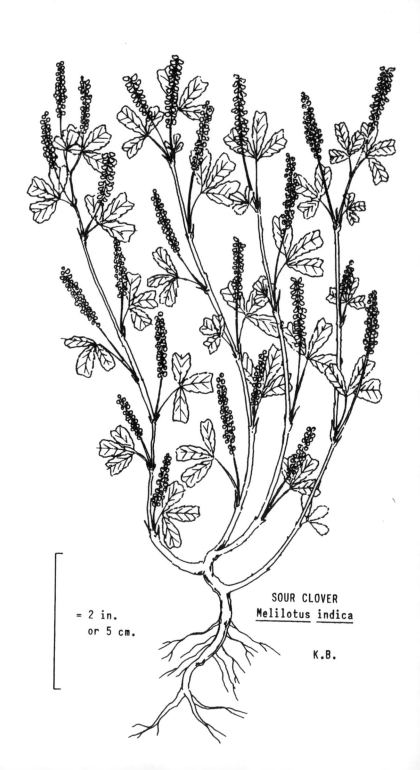

= 2 in.
or 5 cm.

SOUR CLOVER
Melilotus indica

K.B.

SAINFOIN
Onobrychis sativa

Onobrychis <u>sativa</u> (O. <u>viciaefolia</u>, O. <u>viciifolia</u>,
  Heydysanum <u>onobrychis</u>) SAINFOIN, ESPARSETTE

HABIT: A perennial legume. Sainfoin is long lived
  on dry land, but short lived on irrigated land. Its
  roots penetrate 2 to 3 times deeper than Alfalfa roots.

USES: Hay, forage and a good bee plant. Sainfoin is
  the most important fodder plant for dry and calcareous
  hills.

RANGE: Sainfoin is adapted to temperate, rather than
  cold climates. In Europe, its distribution is about the
  same as the grape, although it can grow well in places
  too cool for grapes. Wet and cold summers diminishes
  its yield. This species will withstand severe drought.

SOIL: Sainfoin is best adapted to well drained,
  calcareous soil, as in the northern Rocky Mountain
  region. It does not grow on acid soil. On non-calcar-
  eous soils, Sainfoin grows well if heavily limed.
  This plant also grows well in soils low in phosphorus.

SEEDING: Sow Sainfoin along with Barley. The Barley
  grows as a nurse crop for Sainfoin. It is best to sow
  them at right angles to each other. A recommended
  seeding rate is 30 to 60 lb.s per acre at a depth of
  ½ inch. This plant does not establish readily so do
  not graze it heavily during the first autumn. Hay
  may be cut the next year when the flowers open. The
  seed remains viable for a relatively short time.
  /a9, h5, h6, p4, p6, r1, w3/

Ornithopus <u>sativus</u> SERRADELLA

HABIT: An annual legume.

USES: Forage and green manure.

RANGE: Serradella is native to the Spanish peninsula
  and Morocco. It is adapted to regions with a mild
  winter climate or a cool growing season. The young
  plants will withstand frost, but are killed by a
  temperature of 20°F. This plant will not withstand
  drought.

SOIL: Growth is best in moist sand or sandy loam

soil. Lime is not beneficial, but often deleterious
to Serradella's growth. Actually, it is better that
lime be absent from the soil for the growth of this
crop. A striking characteristic of Serradella is
that it definitely needs inoculation. A recommended
seeding rate is 15 to 20 lb.s per acre of unhulled
seed.
/h6, p4, p5/

Panicum miliaceum PROSO MILLET, BROOM CORN MILLET, HOG
HABIT: A rapidly growing, summer annual grass. It is
  shallow rooted.
USES: Grain and hay – possibly green manure. Proso
  Millet is considered to be more valuable for grain
  than hay. It yields a large quantity of grain which
  is good for feeding hogs and chickens. By comparison,
  Foxtail Millet produces a more desirable hay than Proso.
  Proso is a better catch crop because it matures quickly.
  It matures two weeks earlier than Foxtail.
RANGE: Proso Millet grows best in warm weather with
  plenty of moisture. This species is considered to be
  a better dryland crop than Foxtail Millet. By maturing
  quickly, Proso Millet grows well with little water.
  Nevertheless, it is injured by severe drought. Proso
  may be sown farther north than Foxtail Millet. It has
  been grown in the Dakotas, Montana and Canada.
SOIL: Proso Millet is adapted to a wide range of
  soils. It grows best on well drained, fertile, loam
  soils. Being shallow rooted, Proso Millet is unadapted
  to sandy or droughty soils.
SEEDING: Sow at a rate of 15 to 25 lb.s per acre.
  /a7, h5, h6, w4/

Panicum texanum TEXAS MILLET
HABIT: An annual grass.
USES: Forage.
RANGE: A native of the bottom lands of Texas and
  Mexico.

SOIL: It grows best on clay and loam soils.
   /p4/

Pennisetum typhoides (P. glaucum) PEARL MILLET, MILLET
HABIT: A summer, annual grass with shallow roots.
USES: Grain, hay, pasture, silage, smother crop and
   green manure. This species matures quickly so it is
   frequently utilized as a catch crop. It is a good
   crop for smothering weeds.
RANGE: Pearl Millet is a warm climate crop. The best
   temperature range for this plant is 76 to 86^{0}F with a
   minimum temperature of 59^{0}F. It will mature seed as
   far north as Maryland. The annual rainfall can be as
   low as 16 to 26 inches, but better growth occurs with
   more moisture, such as when irrigated in the Southwest.
   Pearl Millet will go dormant during a drought, then
   resume growth. The elevation limit for this crop is
   6,500 to 8,900 feet in the Southwest.
SOIL: Pearl Millet grows on a wide variety of soils.
   Best growth is produced on fertile, loam soils with
   good drainage. It will grow on poor soil.
SEEDING: A recommended seeding rate is 20 to 30 lb.s
   per acre at a depth of $\frac{1}{2}$ inch. The soil must be
   warm before sowing.
   /a7, a8, h6, m6, s4, w2, w4/

Phaseolus aconitifolius MOTH BEAN, MAT BEAN
HABIT: A summer growing, annual legume. The foliage
   of this plant covers the ground so completely that
   there is practically no water evaporation from the
   soil.
USES: Forage, hay cover crop, green manure, grain
   and vegetable. When young, the pods can be eaten as a
   vegetable. It only needs two to three months to mature
   so it is often grown as a summer catch crop. Moth
   Bean can be used for hay, but it is difficult to harvest
   because the vines lie close to the ground.
RANGE: Moth Bean is well adapted to northern Texas and

California. It is more drought resistant than the Cow
Pea. This is a hot weather legume, thriving where temperatures are high. It is often planted with hot
weather crops like Sorghum or Millet. This plant
requires a remarkably small amount of moisture. If it
is planted at the end of the rainy season, the moisture
remaining in the soil is enough to mature a crop.

SOIL: Moth Beans will grow on a wide variety of soils,
including poor soils. Best growth occurs in sand and
sandy loam soils.

SEEDING: A recommended seeding rate is 35 to 40 lb.s
per acre
/k1, n1, p4, s3/

Phaseolus acutifolius TEPARY BEAN, TEXAS BEAN

HABIT: A rapidly growing, summer annual legume. It
matures in only two months. The ripe seeds have no
dormancy period and can germinate immediately. Thus,
the whole plants should be harvested when the first
pods ripen.

USES: Forage, hay, green manure and grain. However,
the seeds are hard to cook. Tepary Beans serve well as
a catch crop. They mature in two to three months. If
irrigation water is available, two crops can be
produced before cool weather ends the growing season.

RANGE: Originating from Mexico, this species grows in
arid and semiarid regions. It is a drought resistant
plant which can be grown with irrigation or on residual
soil moisture. Tepary Beans often survive in climates
too arid for other beans. This plant needs ample
moisture to germinate and initiate growth, but after
that, little or no rain is needed. A single rainfall
can be enough to complete the growth process if the
soil is deep and retains moisture.

          Tepary Beans grow poorly in humid climates.
They also cannot tolerate waterlogged soils or frost.
In temperate latitudes, flowering is delayed until
short days arrive.

TEPARY BEAN
Phaseolus acutifolius

SOIL: Tepary Beans thrive on light, sandy soil.
SEEDING: Sow at 25 to 30 lb.s per acre. The seed has a
  medium relative longevity.
  /a9, h6, k1, n1, s3, s7/

Phaseolus angularis ADZUKI BEAN
HABIT: A summer annual legume.
USES: Forage and grain. This crop produces a large
  quantity of seed.
RANGE: Adzuki Beans are adapted to the same growing
  conditions as the Soybean - a hot, moist climate.
  This crop needs to be cultivated because it does not
  compete well with weeds.
SOIL: It will grow in almost any soil.
SEEDING: A recommended seeding rate is 20 to 25 lb.s
  per acre when spaced in rows 3 to 4 feet apart.
  /p4, s3/

Phaseolus aureus MUNG BEAN, GOLDEN GRAM
HABIT: A summer annual legume.
USES: Forage, grain and green manure. Poultry are
  particularly fond of the seed.
RANGE: Mung Bean grows in the same regions as the Cow
  Pea. It grows best in a hot, moist climate, but also
  grows well in drier climates. Temperatures can range
  from 68 to 113°F.
SOIL: It will grow in any soil, even heavy soils or
  soils of low fertility. This crop has a low lime
  requirement.
SEEDING: Sow Mung Beans only after the soil is warm. A
  recommended seeding rate is 60 to 70 lb.s per acre at
  a depth of 1 inch. The seed remains viable for a
  relatively short time.
  /a5, h6, k1, p4, s3, s8/

Phaseolus limensis (P. lunatus) LIMA BEAN, BUTTER BEAN,
  FRIJOLITA DE CUBA
HABIT: A summer annual legume. Lima Beans need a

AA

1CM.

MUNG BEAN
Phaseolus aureus

long growing season — five to seven months to mature.
The pole varieties are later maturing than the dwarf.
Two to three cultivations are necessary when the plants
are young. After this, the vines cover the ground.
USES: Green manure, cover crop, grain and, when young,
the pods and leaves can be used as a vegetable.
RANGE: Lima Beans grow best in a warm, sunny climate
with a high relative humidity; such as exists along
the Southern coast of California. They originated
in the American tropics. This crop is very suseptible
to frost damage. Once established, they are highly
resistant to drought, although less so than Kidney
Beans. Limas grow at elevations up to 7,900 feet.
SOIL: Soil is a less limiting factor than climate for
growing Lima Beans. They grow well on alluvial soils
and on poor soils. Well drained, well aerated soils
with a pH between 6 and 7 are best for this crop. Lima
Beans will withstand more alkaline soil conditions than
varieties of Phaseolus vulgaris.
/h1, h2, n1, s7/

Phaseolus mungo GREEN GRAM
Phaseolus mungo var. radiatus BLACK GRAM, URD BEAN
HABIT: Summer annual legumes.
USES: Grain and green manure. In India, Black Gram
is often interplanted with Cotton or Sorghum. It is
also grown in rotation following Rice.
RANGE: Black Gram can be grown in dry areas with low
rainfall. Temperatures can range between 68 and 113°F.
/c1, s3, s7, s8/

Phaseolus vulgaris KIDNEY BEAN, HARICOT BEAN, COMMON
FIELD BEAN
HABIT: A summer annual legume.
USES: Forage, grain and green manure. Kidney Beans
can be rotated between two successive Cereal crops.
They can also be interplanted with Cotton or Corn —
alternating two rows of each. Some other rotations

are:

>   1) Kidney Beans, Wheat and Clover – all
>   grown one year.
>   2) Corn or Potatoes, Kidney Beans, Wheat
>   and Clover.

RANGE: High temperatures and plenty of moisture are
required by Kidney Beans for germination and growth.
Southern Michigan and Central New York have climates
well suited for this crop. The Pink, Pinto and Red
Mexican varieties of Phaseolus vulgaris are more heat
and drought resistant than the Kidney varieties.

SOIL: Kidney Beans are influenced more by climate
than soil type. This legume will grow on poor soil
or any soil ranging from clays to sand. They grow
best on well drained loam or calcareous soils. This
plant does not tolerate waterlogged or saline soils.

SEEDING: Sow Kidney Beans only after the soil is warm.
A recommended spacing is 16 to 20 inches between rows
and 3 to 5 inches between plants. When sowing on sandy
or gravelly soil during dry weather, the seeds should
be soaked in soft water for several hours. Then roll
the seeds in ground plaster. This will help insure
germination.
/h1, h5, s7, s8, w3/

Phleum pratense TIMOTHY

HABIT: A short lived, perennial bunchgrass which
reproduces by seed. It does not spread vegetatively
or form a sod.

USES: Forage and hay. In the U.S., Timothy is
considered to be the most important hay grass.

RANGE: Timothy is a northern grass. It does not
succeed well south of 36° latitude, except at higher
elevations. This species is best adapted to cool,
moist climates, as found in the coastal region of the
Pacific Northwest, and the Northeast and Northcentral
states east of the Missouri river. It is more cold
resistant than most cultivated grasses. Timothy does

TIMOTHY
Phleum pratense

not withstand hot, humid, summer weather. A lack of
water is damaging to Timothy, especially when producing
its seed.

SOIL: Timothy is adapted to a wide range of soil
types. Best growth occurs on clay or loam soils.
This plant needs good drainage. Nevertheless, it
thrives best where moisture is abundant. An application of lime helps Timothy grow well on acid soils.

SEEDING: Sow at a rate of 6 to 12 lb.s per acre.
/a7, h5, h6, p4, p6, w4/

Pisum arvense (P. sativum subsp. arvense) FIELD PEA,
CANADIAN FIELD PEA, AUSTRIAN WINTER PEA

HABIT: An annual legume which grows in the winter
where the climate is mild or in the spring where the
winters are too severe for growth. When planted once
every three years, damage from disease is greatly
reduced.

USES: Forage, hay, silage, grain and green manure.
Peas and Oats are one of the best mixtures for hay.
When Field Peas grow vigorously, they outcompete weeds,
helping to reduce their population.

RANGE: Field Peas grow in all parts of the U.S. and
Canada. They require cool, moist growing conditions.
This crop will withstand heavy frost. However, they
quickly succumb to heat, especially if combined with
humidity. Heat is particularly disasterous to seed
production.

SOIL: Most soil types are suitable for Field Peas,
provided they are well drained. On sand, Peas can
quickly suffer from drought because they are surface
rooters. Field Peas will produce on any soil that will
produce Oats. Growth is best on clay loams, loams or
sandy loams. They also grow well on calcareous soils,
preferring an abundance of lime. When Field Peas are
grown on heavy, black soils rich in humus, they produce
a heavy growth of vines with few pods.

SEEDING: A recommended seeding rate is 70 to 90 lb.s

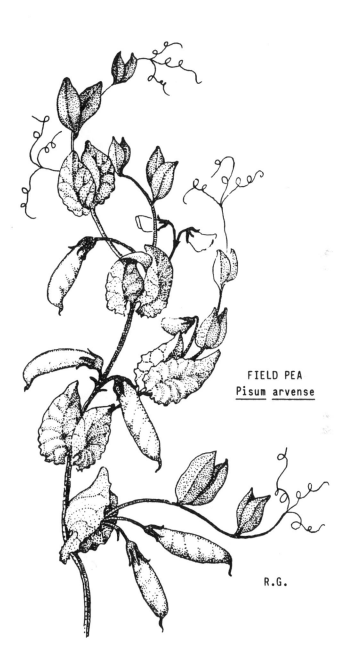

FIELD PEA
Pisum arvense

R.G.

per acre at a depth of 1½ inches. Sow the seed thickly
when broadcast because when thinly sown, the plants will
fall down and rot. Another way to prevent this problem
is to sow a nurse crop. Oats are the best nurse crop
for Field Peas. Innoculation is essential for good
Field Pea growth. The seed remains viable for a relatively short time. If the seeds are buggy, immerse
them in boiling water for two minutes to kill the Pea
bug. This short interval will not harm the seed.
/a5, a7, a9, h5, h6, m4, p4, p5, p6, r1, w3, w4/

Poa compressa CANADA BLUEGRASS, VIRGINIA BLUEGRASS, WIREGRASS

HABIT: A long lived, perennial grass.
USES: Pasture and hay. Canada Bluegrass does not
  yield heavily when cut for hay. It is good as a
  pasture plant because it will withstand close grazing.
RANGE: Canada Bluegrass is adapted to a cool, moist,
  temperate region - the same as Kentucky Bluegrass. It
  is more resistant to summer heat and drought than
  Kentucky Bluegrass.
SOIL: Canada Bluegrass grows best in the same soils
  as Kentucky Bluegrass - well drained, loams or clay
  loams rich in humus and clacium. But, Kentucky
  Bluegrass grows better in shade, or wet soils, than
  Canada Bluegrass. On the other hand, Canada Bluegrass
  grows better on stiff clays and thin, gravelly soils
  where Kentucky Bluegrass won't grow. Canada Bluegrass
  is dominant only on poor soils, too acid, droughty or
  deficient in nutrients, for Kentucky Bluegrass or other
  grasses to predominate.
SEEDING: Sow at a rate of 15 to 25 lb.s per acre.
  /a7, h6, p4, w4/

Poa pratensis KENTUCKY BLUEGRASS
HABIT: A long lived, sod forming, perennial grass
USES: Forage. Kentucky Bluegrass is one of the most
  important pasture grasses. It is frequently grown with

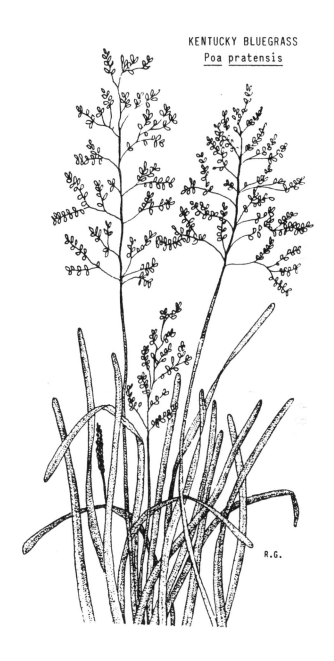

KENTUCKY BLUEGRASS
Poa pratensis

R.G.

White Clover. Growth begins earlier in the spring than
most grasses. This grass makes good pasturage in the
spring and fall.

RANGE: Kentucky Bluegrass is widely distributed
throughout ths U.S. and Canada. It is adapted to cool,
moist, temperate regions with high humidities. In arid
regions, it succeeds well with irrigation. Cold does
not affect Kentucky Bluegrass, although this plant
languishes during summer heat. Most vigorous growth
occurs in temperatures between 60 to 80^{0}F. Kentucky
Bluegrass is easily affected by drought because it
has shallow roots. Nevertheless, dry weather does
not kill this plant. It survives droughts by going
dormant. This is a sun loving plant. During the hot
summer, it grows well in semi-shade, for example, in
open woodlands.

SOIL: Well drained loams or clay loams, particularly
those rich in humus, are best for Kentucky Bluegrass.
The most successful Bluegrass growing soils are those
developed from calcareous material. Along with calcium,
nitrogen and phosphorus are important for its good
growth. Kentucky Bluegrass grows best in soil with a
pH between 6.0 and 7.0.

SEEDING: A recommended seeding rate is 15 to 25 lb.s
per acre.
/a7, h5, h6, p4, p6, w4/

Pueraria hirsuta (P. lobata, P. thunbergiana) KUDZU

HABIT: A deep rooted, rankly growing, perennial legume.
During the mid summer, Kudzu makes very rapid growth.
It often becomes a pernicious weed. Watch out!

USES: Cover crop, hay, pasture and vegetable. Kudzu
is one of the best perennial legumes for controlling
erosion on banks and gullies. The starchy roots of
this plant can be eaten.

RANGE: Kudzu was introduced from Japan. It is adapted
to the same moist, hot regions as Lespedeza. This plant
is drought resistant.

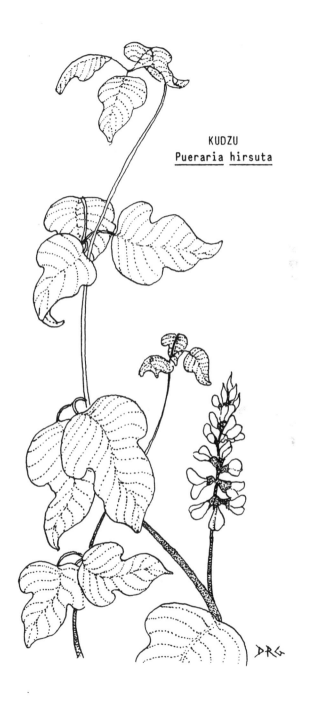

KUDZU
Pueraria hirsuta

DRG

SOIL: Well drained loams or heavy clays produce the best growth. Where the soil is worn out and eroded, Kudzu is invaluable because its growth conserves and improves them.

SEEDING: Kudzu is propogated by root cuttings set 10 feet apart.

/a7, h1, h6, m3, m6, p4, p6/

Secale cereale RYE, CEREAL RYE

HABIT: An annual grass which germinates and grows rapidly.

USES: Grain, hay, pasture, cover crop and green manure. Rye is one of the best crops where fertility is low and winter temperatures are extreme. Rye should be used only in rotation with row crops because other grain crops are graded down in the market if they contain Rye seed. Rye is a good green manure because it produces large quantities of organic matter. It is a better soil renovator than Oats.

RANGE: Rye is a native of Southwest Asia and the Mediterranean region. This crop has a wide range of adaptability. Being very winter hardy, it is seldom injured by cold weather. No other winter cereal can be grown as far north as Rye. This species is hardier than Winter Wheat, grows better in cooler weather and can be turned under earlier in the spring.

          Rye grows best with ample moisture. However, in low rainfall regions, where legume cover crops seldom make much growth, Rye grows well.

SOIL: This grass will grow on any soil, even in acid soils, heavy clays, light sands, and infertile or poorly drained soils. Rye will grow on soils too poor to produce other grains or Clover. Best growth occurs on well drained loam or clay loam soils. It has a low lime requirement.

SEEDING: A recommended seeding rate for green manure is 90 to 160 lb.s per acre at a depth of 3/4 inches. The seed remains viable for a relatively long time.

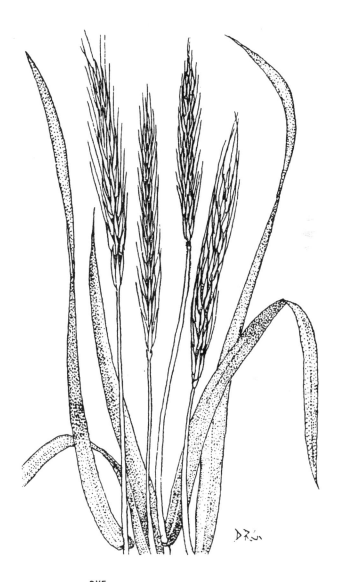

RYE
Secale cereale

/a5, a7, b2, h6, m3, m7, m10, p2, s4, w2, w3/

Sesbania macrocarpa (S. exaltata) COMMON SESBANIA, HEMP SESBANIA

HABIT: A summer annual legume.

USES: Green manure. The roots of Sesbania open up the subsoil.

RANGE: Sesbania is a subtropical plant which grows best in a climate with high temperatures, low relative humidities and moist soils. It does exceptionally well in the Southwest where these conditions are met. This legume grows well with irrigation. Considerable drought and salinity can be endured. Sesbania will grow at higher altitudes than Crotolaria.

SOIL: Sesbania will grow on any soil, even soils of low fertility. Best growth occurs on rich, loam soils. This species has a low lime requirement.

SEEDING: A recommended seeding rate is 20 to 25 lb.s per acre at a depth of 3/4 inches. The seed remains viable for a relatively long time.

/a5, h6, m2, m6, p5/

Setaria italica FOXTAIL MILLET, ITALIAN MILLET

HABIT: A rapidly growing, summer annual grass. It matures faster than Sorghum.

USES: Grain, pasture, hay, silage, smother crop and green manure. Foxtail Millet matures quickly. It fits well in rotations as a catch crop. Its rapid growth makes it valuable for subdueing weeds. With proper heat and moisture, a grain crop can be harvested in six to ten weeks.

This crop is a good feed for cattle, but is injurious to horses, causing derangements in the kidneys, swelling the joints and softening the bones. As a feed, it is best when cut in early bloom.

Foxtail Millet is often planted on newly turned sod. This practice aids in disintegrating the sod. It must be noted that Foxtail Millet has a

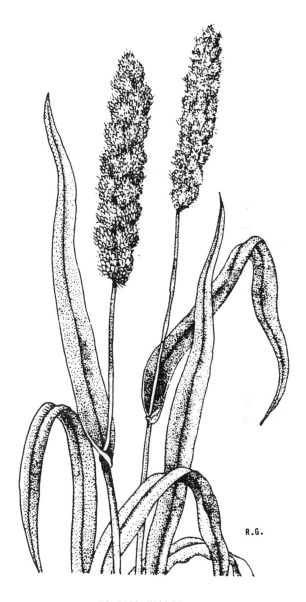

FOXTAIL MILLET
Setaria italica

reputation of reducing the yield of the following crop.

RANGE: Foxtail Millet requires the same hot, moist
climate as Sorghum. It is well adapted to the Northern
and Central Great Plains states. Growth is best in
regions with fairly abundant rainfall. This plant is
able to grow with small supplies of moisture. Therefore, it is useful in dry regions. Foxtail Millet is
drought resistant, but less so than Sorghum. If injured
by a prolonged drought, Foxtail Millet lacks the
ability to recover.

SOIL: Growth is best on sandy loam soils. It will
grow on poor soils. Good drainage is essential for
proper growth. This grass has a low lime requirement.

SEEDING: Sow only after the soil is warm and continued
warm weather is assured. A recommended seeding rate
is 20 to 30 lb.s per acre at a depth of $\frac{1}{2}$ inch.   The
seed remains viable for a relatively long time.
/a5, h5, h6, p3, p4, p6, w4/

Sorghum bicolor (S. vulgare) SORGHUM

HABIT: A summer growing, annual grass. Sorghum
continues to grow later in the fall than Corn. Its
climatic and soil adaptations are almost identical
with Corn.

USES: Grain, animal feed and green manure. In
semiarid regions, Sorghum serves the same purpose as
Corn. It is grown instead of Corn because it is more
tolerant of heat, low humidity and drought. Sorghum is
frequently grown with Cow Peas. Sorghums stalks and
roots take longer to decompose than Corns'. Reportedly,
Sorghum reduces the yield of succeeding crops.

RANGE: Being a native of tropical Africa and Asia,
Sorghum grows best in hot, sub-humid climates. It
is injured by light frosts. Sorghum can be grown in all
parts of the U.S., almost up to the northern boundary.
Best growth occurs in the southern half of the U.S.,
where temperatues are uniformly high during the growing
season. The best average temperature is 80°F.

This species will grow at altitudes up to 5,000
to 6,000 feet in Wyoming and Colorado, and 7,000 feet
in Arizona and New Mexico. Sorghum grows best with
irrigation or where rainfall is abundant. However,
it can be grown where the annual rainfall is as low
as 16 to 26 inches. In regions of uncertain rainfall,
Sorghum is valuable because it resists wilting and goes
dormant during a drought. It resumes growth when
sufficient moisture is available.

SOIL: Sorghum grows on most soils, provided they are
well drained and have a permeable subsoil. Soil too
poor and thin for Corn or Wheat seldom produce satisfactory Sorghum crops. Sandy loam soils are best for
Sorghum. It is more tolerant of alkaline soils than
most crops.

SEEDING: A recommended seeding rate is 90 lb.s per acre
at a depth of 3/4 inches. Sow after the soil is warm
and air temperatures are consistently high. The seed
remains viable for a relatively long time.
/a5, a7, h6, p4, s4, s5, w2, w4/

Sorghum bicolor var. sudanense (S. vulgare var. sudanense)
SUDANGRASS

HABIT: A rapidly growing, summer annual grass.

USES: Hay, pasture and green manure. Sudangrass is
relished by cattle and horses. This crop requires a
short growing season. Consequently, it is valuable as
a catch crop. Sudangrass matures hay in 75 to 80 days
and seed in 100 to 106 days when the weather is hot.

RANGE: Sudangrass is adapted to the same hot and moist
conditions as Sorghum, although it ripens earlier and
will mature as far north as 49° latitude. Sudangrass
grows in nearly every part of the U.S.. Areas where it
cannot be grown successfully are those which have cool
weather or light frosts. Excessive humidity is also
injurious to its growth. These conditions interfere
with its normal development. It will also not grow at
high altitudes. Sudangrass endures high temperatures

SUDANGRASS
Sorghum bicolor var. sudanense

and drought. When grown in rows with cultivation, it
can be grown with very little water.
SOIL: Sudangrass will grow on any soil, except those
  that are cold and wet. Alkaline soils also reduce its
  yield. Sudangrass grows best on loams or clay loams.
  It has a low lime requirement.
SEEDING: Sow in the late spring or early summer, after
  the soil is warm and the weather settled. A recommended
  seeding rate is 35 lb.s per acre at a depth of 3/4
  inches. The seed remains viable for a relatively long
  time.
  /a5, a7, m2, m6, p3, p6, w4/

Spergula arvensis SPURRY
HABIT: A rapidly growing, annual, broadleaf plant.
USES: Hay, pasture and green manure. This plant
  produces a mature crop, 12 to 14 inches high, in 7
  to 8 weeks. Three crops may be grown in one season.
  For example, when sown in March, May and July, Spurry
  produces three crops that renovates a poor soil so
  it will produce Clover or winter grain. Spurry is
  called the "Clover of sandy land".
RANGE: Spurry is adapted to a cool, moist growing
  season. The young plants do not withstand heat or
  drought. When older, frosts are not injurious.
SOIL: Sandy soils are best for Spurry. It will grow
  on dry and sandy land where Clover fails.
  /p4, w3/

Spergula maxima GIANT SPURRY
          Giant Spurry matures later than Spurry. It
  is adapted to heavier soils rich in lime.
  /p4/

Stizolobium deeringianum VELVET BEAN, DEERING or FLORIDA
  VELVET BEAN
HABIT: A summer growing, annual legume. Velvet Bean
  is the most vigorous annual legume planted in the U.S..

= 2 in.
or 5 cm.

SPURRY
Spergula arvensis

K.B.

USES: Livestock feed, grain and green manure. This
  legume yields heavily and decays readily. Velvet Bean
  has been used for green manure in Florida Citrus. It
  is often interplanted with Corn. The Corn acts as a
  supporting crop.

RANGE: Velvet Bean is a semi-tropical plant. Its
  range is limited because it requires a long season of
  hot weather. A light frost destroys this legume. It
  can be grown south of the Ohio river, throughout the
  Cotton belt, and on the well drained soils of the
  South Atlantic and Gulf states.

SOIL: This species will grow on any soil provided it
  is well drained. It grows well on soils of low fertility. Best growth occurs on loam soils. Velvet Bean
  has a low lime requirement. For poor sandy soils,
  Velvet Bean is the most vigorously growing legume
  known. In this soil situation, applications of phosphate and potash are beneficial.

SEEDING: Sow after the soil is warm. A recommended
  seeding rate is 120 lb.s per acre at a depth of 2
  inches. The seed remains viable for a relatively short
  time.
  /a5, h1, h5, m2, m6, p4, p5, w4/

Trifolium alexandrinum BERSEEM CLOVER, EGYPTIAN CLOVER

HABIT: A summer annual legume. In Southwestern
  rotations, it grows as a winter crop.

USES: Forage and green manure.

RANGE: Berseem Clover grows in most of the U.S.. This
  Clover does not withstand extreme hot or cold temperatures. It is the least winter hardy of the cultivated
  Clovers. Berseem Clover grows well along the Gulf
  coast, and in the Yuma, Rio Grande and Imperial valleys
  of the Southwest.

SOIL: Berseem Clover tolerates alkaline soils. It
  thrives under irrigation.

SEEDING: Sow at 15 to 20 lb.s per acre.
  /h6, L2, k1, p4, p5/

Trifolium dubium SMALLHOP CLOVER
Trifolium procumbens BIGHOP CLOVER
Trifolium agrarium FIELDHOP CLOVER
HABIT: Winter annual legumes.
USES: Pasture. Fieldhop Clover is the least useful
  of the three species.
RANGE: These Clovers are cool season plants which are
  more prevalent at high elevations. They grow in the
  southern half of the U.S., the Atlantic coast states,
  and westward in Kentucky, Missouri, Oklahoma and the
  Pacific coast states. Smallhop Clover germinates and
  establishes well under dry conditions.
SOIL: They will grow on any soil, even when infertile
  and eroded.
SEEDING: A firm seedbed is necessary for planting. Sow
  at a rate of 3 to 5 lb.s per acre. Once established,
  all three continue to persist and spread. The seed
  remains viable for a relatively long time.
  /a9, h6, p4, p6, r1, w4/

Trifolium fragiferum STRAWBERRY CLOVER
HABIT: A summer growing, perennial legume. Common
  Strawberry Clover is self fertile. The Salina variety
  is self sterile and needs cross pollination.
USES: Pasture, cover crop and green manure.
RANGE: Strawberry Clover is a warm season plant, yet
  it will persist in a climate with extreme hot or cold
  weather. Salina is not as winter hardy as Common.
  Common will even withstand snow cover.
SOIL: Strawberry·Clover is adapted to a wide range of
  soils. It grows best where moisture is plentiful. One
  to two months of flooding are tolerated by this Clover.
  Strawberry Clover even grows on the salty soils along
  tidal streams. Due to its ability to grow in wet
  conditions, this plant has been useful in improving
  swampy ground. It is valuable in irrigation projects
  where drainage is a limiting factor. Strawberry Clover
  also endures dry, alkaline soil.

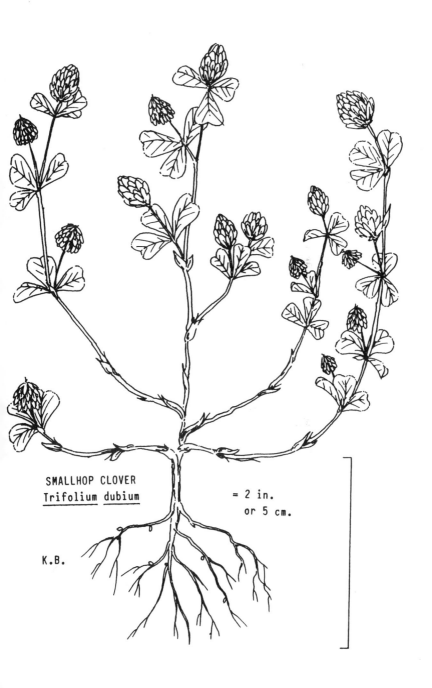

SMALLHOP CLOVER
Trifolium dubium

= 2 in.
or 5 cm.

K.B.

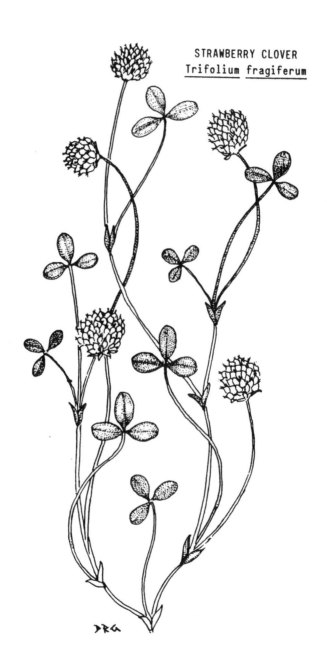

STRAWBERRY CLOVER
Trifolium fragiferum

The Salina variety of Strawberry Clover is
more tolerant of poorly drained soil conditions,
drought and salinity than Ladino Clover. However, it
produces best on fertile, well drained soil. Salina's
roots will descend to 2 to 4 feet deep when there isn't
any subsurface hardpan. When the soil is porous,
Salina requires fewer irrigations than Ladino. When
grown without irrigation in California, in regions
with 36 to 40 inches of annual rainfall, Salina
produces continuously throughout the entire growing
season, including a rain-free period of 90 to 100 days.
        Strawberry Clover is more suseptible to
Sclerotinia disease than Ladino Clover. Sclerotinia
is favored by cool, wet conditions.
SEEDING: Early fall sowing, before cold weather sets in,
    germinate better than spring sowings. However, weed
    control in the spring may be easier because the weeds
    can be mowed at the proper time. With fall sowings,
    the soil is generally too moist to mow at the proper
    time so weed control is delayed. A recommended seeding
    rate is 6 to 10 lb.s per acre. The seed remains viable
    for a relatively long time.
    /a9, h6, L2, p1, p4, p6, r1, w4/

Trifolium hirtum ROSE CLOVER
HABIT: A winter annual legume. No other introduced
    legume is as well adapted to such a wide variety of
    climatic conditions or soil types as Rose Clover.
USES: Forage, cover crop, erosion control and green
    manure. Rose Clover will grow on poor soils where
    few other plants survive. It is best to grow Rose
    Clover alone because it is a poor competitor with
    other species. This Clover grows well when mowed.
RANGE: Rose Clover is a native of the Mediterranean
    region. In the U.S., it grows in the Southern states
    and California. Rose Clover will survive under
    adverse conditions. This species can grow in areas
    receiving as little as 10 inches of annual rainfall,

ROSE CLOVER
Trifolium hirtum

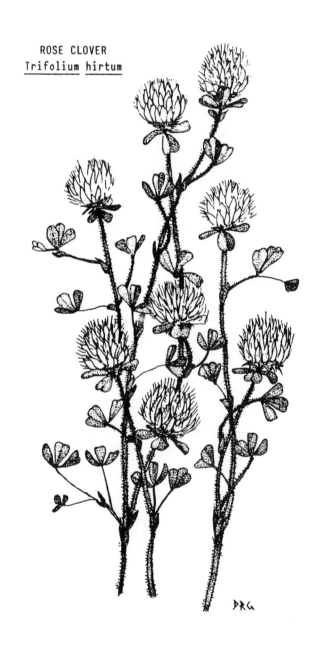

or at elevations up to 3,300 feet. It is a winter
hardy plant. Late spring frosts usually do not injure
Rose Clover.

SOIL: Rose Clover will grow in any soil provided it
is well drained.

SEEDING: A recommended seeding rate is 15 to 20 lb.s
per acre. The seed remains viable for a relatively
long time. This legume reseeds itself well.
/a9, b2, g1, h6, L2, m11, p6/

Trifolium hybridum ALSIKE CLOVER

HABIT: A perennial legume which lasts three to five
years. It is often treated as a biennial. Alsike
Clover is adapted to a wider range of climatic and
soil conditions than Red Clover, and nearly as great
as White Clover. Its root system is closer to the
soil surface than Red Clover's.

USES: Green manure, pasture and hay. It is also an
excellent bee plant.

RANGE: Alsike Clover originates from Sweden. It
thrives in cool climates with abundant moisture – more
than 38 inches of annual rainfall. This Clover endures
cold and heat better than Red Clover and perhaps
drought. It rarely winter kills. In the U.S., Alsike
Clover grows north of the Ohio and Potomac rivers,
west to the Minnesota–Dakota boundary, in Idaho and
along the Washington and Oregon coast. In the South,
it is less successful because it usually disappears
during the hot summer months.

SOIL: Alsike Clover will grow on any soil provided it
has abundant moisture. This plant thrives on heavy
silt soils, peat soils, clays, clay loams, sandy loams
and muck soils. It has a medium lime requirement and
responds well to an application of lime. Timothy grows
well with Alsike Clover, although Alsike will grow in
soils with far less clay in them.

          Alsike Clover is less suseptible to pests than
Red Clover. It thrives where Red Clover dwindles due

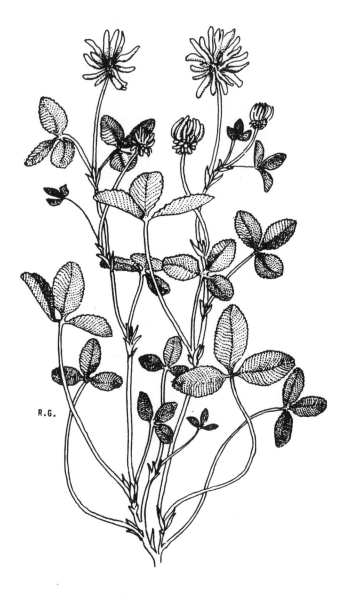

R.G.

ALSIKE CLOVER
Trifolium hybridum

to Clover failure. This permits almost continuous
growth on the same land.
SEEDING: Sow in either spring or fall. A recommended
seeding rate is 6 to 8 lb.s per acre at a depth of
$\frac{1}{2}$ inch. If heavily seeded at 15 lb.s per acre, it
makes a dense, heavy cover 2 feet deep. The seed
remains viable for a relatively long time.
/a5, a7, a8, a9, b1, h6, L2, p4, p5, r1, s5, w4/

Trifolium incarnatum CRIMSON CLOVER
HABIT: A summer annual legume. Crimson Clover grows
during the winter in mild climates. It usually cannot
grow in the winter north of New Jersey.
USES: Hay, pasture, cover crop and green manure. When
cut before bloom, Crimson Clover makes a good hay. But,
you should note that it is a hairy plant and tends to
form hair balls in animals' stomachs. In northern
areas, Crimson Clover can be spring sown and grown as
a summer crop. This is done for rotation with potatoes
in Maine. For fall sowing, frequent late summer and
early fall rainfalls are required so the plants can
get a good start. Adequate soil moisture is essential
for good germination. As a green manure, Crimson
Clover can be winter grown in rotation between Milo
and Cotton, for example; or Soybeans, small grain
and Cotton. Crimson Clover grows reasonably well in the
fall and rapidly in the spring so it will attain
maximum development by the time it must be turned under
for Corn or any other summer crops. It decomposes
rapidly when turned under.
RANGE: Crimson Clover grows well in cool, humid
weather with 35 inches or more of annual rainfall. It
requires considerable heat in the early stages of
growth. It cannot endure much freezing, nor extreme
heat and drought. Seedlings which have sprouted well
can be killed by subsequent dry weather. After the
seedlings become well established, Crimson Clover
makes more growth than most other Clover species.

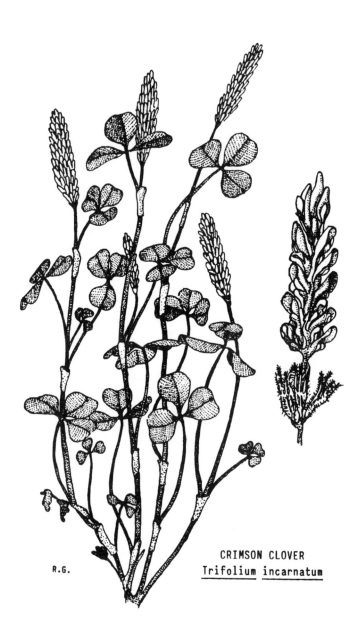

R.G.

CRIMSON CLOVER
Trifolium incarnatum

Crimson Clover is principally cultivated in the south
Atlantic states, especially in the states north of
the Cotton belt.

SOIL: This legume will grow on almost any soil,
provided it is well drained. Muck soils or extremely
acid soils do not support Crimson Clover. Best growth
occurs on loam soils with a good humus content.
Crimson Clover is adapted to soils of low fertility and
has a medium lime requirement. On low fertility soils,
phosphate and potash fertilizers and manure will help
you obtain better stands. Another method to prepare
poor soil is to turn under two Cow Pea crops before
sowing a Crimson Clover crop.

        Crimson Clover is adapted to withstand shade.
It is often sown in orchards and with other crops.
Growth is best when it is mowed. This legume has been
commonly sown with Corn or other cultivated crops,
following the last cultivation, either just before or
just after a soaking rain.

SEEDING: Like other Clovers, Crimson prefers a firm
seedbed. Shallow drilling produces a better stand than
broadcasting. A recommended seeding rate is 15 to 30
lb.s per acre at a depth of $\frac{1}{2}$ inch. If the seed is
unhulled, sow 40 to 50 lb.s per acre. Natural, unhulled
seed germinates better than hulled. This species
reseeds itself well, when allowed to mature. The seed
has a medium relative longevity.
/a5, a7, a8, a9, b1, b2, h5, h6, L2, L4, m3, m4, m11,
p2, p4, p5, r1, w4/.

Trifolium nigrescens BALL CLOVER
HABIT: A legume. It produces most of its growth one
month later than Crimson Clover.
USES: Forage.
RANGE: The lower southeastern U.S..
SOIL: Loam or clay soils
SEEDING: Sow at a rate of 2 to 4 lb.s per acre.
/h6/

Trifolium <u>pannonicum</u> HUNGARIAN CLOVER
        Hungarian Clover is a deep rooted, long lived,
    perennial legume utilized for forage.
    /p4/

Trifolium <u>pratense</u> RED CLOVER
HABIT: A summer growing, biennial or short lived
    perennial legume. Most of its roots are in the top 12
    inches of soil. In the South, Red Clover is used as a
    winter annual. It is sown in the fall and, after
    cutting in spring, few plants are able to survive the
    summer heat. In the North, it is sown in the spring.
        The Mammoth variety (<u>T</u>. <u>pratense</u> var. perenne)
    is adapted to similar soil and climatic conditions to
    those of Red Clover. Common Red Clover differs from
    Mammoth in that Mammoth is more perennial, matures
    later (the same time as Timothy), is more adaptable
    and is a better green manure, but an inferior hay.
USES: Green manure, hay, pasture and an orchard
    cover crop. Red Clover is a very hardy plant which
    displaces weeds. This legume is bumblebee pollinated
    and is a poor honeybee plant.
        A recommended way to treat a two year Red
    Clover crop is to cut the first year growth for hay,
    or, even better, let it fall and decay. Then turn
    under the second year growth for green manure; pr
    harvest the seed crop first, then till in the straw
    aftermath.
        Some of the traditional rotations with Red
    Clover in the Corn belt are: Corn, Soybeans, Wheat
    and Red Clover; or Corn, Oats, Wheat and Red Clover.
    A Potatoes, Wheat and Red Clover rotation has been
    used on Potato farms. Red Clover can also be rotated
    with Milo and Cotton. Wheat and Corn grown after Red
    Clover are generally more free from disease, larger in
    yield and better in quality than those grown with
    animal manure.
RANGE: Red Clover is a temperate climate plant. This

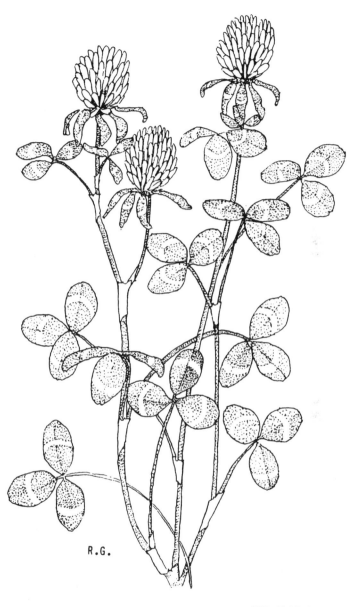

RED CLOVER
Trifolium pratense

is a crop for humid regions without excessive summer or winter temperatures. It will not stand as much cold and moisture as Alsike Clover, nor as much heat and drought as Alfalfa. Red Clover is best adapted to the areas from the Atlantic coast to the 97th meridian and from Canada on the north to the Tennesee-Georgia line on the south. In the Pacific Northwest, this species is valuable under irrigation.

SOIL: Red Clover grows on a wide range of soil types and soils of all states of fertility, except the poorest. Best growth occurs on well drained, clay soils with an abundance of lime. It is intolerant of water logged soil. Conversely, Red Clover will not thrive in gravelly or sandy soils which become droughty. Like Alfalfa, this legume is well adapted to the volcanic ash soils of the West.

Alkaline and saline soil conditions are fairly well tolerated by Red Clover. It will not succeed on land deficient in lime or in soil with a pH as low as 5.5. This species responds well to phosphate fertilization. In shady places, Red Clover does not thrive and generally disappears after the first season.

SEEDING: A recommended seeding rate is 8 to 20 lb.s per acre. Rec Clover can be sown at any time from early spring to early autumn. Sow in a well prepared seedbed at a depth of no more than 1 inch in clay soils and $1\frac{1}{2}$ inches in sandy soils.

Other species, such as Alsike Clover, Alfalfa, Timothy and Bromegrass, are often sown with Red Clover. Red Clover grows especially well with a Ryegrass nurse crop. This legume can also be fall or spring sown in a winter grain crop. Red Clover grows better with a nurse crop than Alfalfa.

Red Clover should not be sown more often than once every eight years due to a fungous disease, Sclerotinia trifoliosum. The land can be alternated with other legume crops, such as other Clovers and Bird'sfoot Trefoil. The seed remains viable for a

relatively long time.
/a7, a9, b1, h5, h6, L2, L5, m2, m6, p4, p5, p6, r1,
s5, w3, w4/

Trifolium repens WHITE CLOVER

HABIT: A hardy, self propogating, long lived, perennial
   legume. White Clover has creeping stems which root at
   the nodes. Its fibrous root system helps it withstand
   drought. This Clover spreads rapidly once it is
   established.

USES: Erosion control, green manure, cover crop and
   pasture. White Clover is not used for hay because it
   is too short and shrinks too much in drying. It is
   honeybee pollinated and a good bee plant. For pastures,
   White Clover grows well with Kentucky Bluegrass, just
   as Red Clover grows well with Timothy. However, it is
   important to note that cattle avoid the flowers and
   when in seed it sometimes causes horses to salivate.

RANGE: White Clover grows in almost the whole temperate
   zone, occurring northward to the limits of agriculture
   and south to the Gulf of Mexico. This Clover thrives
   in cool, moist climates with 35 inches or more of
   annual rainfall, or in the cool part of the year on
   lands that are retentive of moisture. It tolerates
   excessive moisture. White Clover endures cold weather
   better than Red Clover.

SOIL: The abundant growth of White Clover indicates
   a productive soil. Best growth occurs in well drained
   clay, clay loam and loam soils, all rich in humus.
   This plant grows well with plenty of lime, phosphate
   and potash. White Clover is adversely affected by
   chemical nitrogen. Saline or highly alkaline soils
   are not tolerated by White Clover. In eastern Canada,
   an acid tolerant White Clover has been found. White
   Clover grows well in shade.

SEEDING: Sow White Clover early so it becomes well
   established before warm weather arrives. A recommended
   seeding rate is 10 to 12 lb.s per acre at a depth of

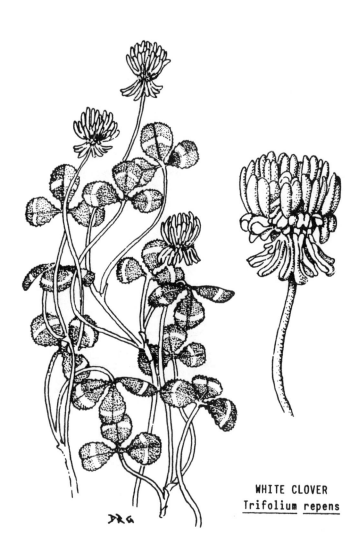

WHITE CLOVER
Trifolium repens

½ inch. White Clover can be broadcast after the last
cultivation of a Corn crop. The seed remains viable
for a relatively long time.
/a8, a9, b1, h5, h6, p4, p6, r1, w3, w4/

Trifolium repens var. latum LADINO CLOVER
HABIT: A medium lived, perennial legume. Ladino
  Clover is a variety of White Clover. This plant is
  low growing with prostrate stems, although it grows to
  twice the size of White clover on irrigated lands.
USES: Green manure, pasture and hay. Ladino Clover
  makes a better green manure than White Clover. It is
  best adapted to rotational pasture management. When
  cut for hay this Clover revives quickly because the
  stems are not cut. Four to five cuttings per year,
  at 35 to 40 day intervals, can be harvested in some
  areas.
RANGE: Ladino Clover requires about the same climatic
  and soil conditions as White Clover. A cool climate
  with a good supply of rainfall is best. Hot, dry
  periods during the summer reduce its growth, but
  usually do not destroy the plants. This variety is
  grown in states with relatively cool, summer temperatures. It is less cold resistant than White Clover.
  Ladino Clover is about as hardy as Red Clover or
  Alfalfa.
SOIL: Ladino Clover is grown primarily on heavy lands.
  Heavy, moist, well drained soils of high fertility are
  best suited for this Clover. It will grow on sandy
  soils during seasons of abundant rainfall. Under these
  conditions, it is not reliable. Ladino Clover does
  not grow well in alkaline soil. The best pH range is
  6.0 to 6.5. It is more tolerant of wet soils than
  Red Clover or Alfalfa, but less so than Alsike Clover.
  /a7, a8, b1, L2, p4, w4/

Trifolium resupinatum (T. suaveolens) PERSIAN CLOVER
HABIT: A winter annual legume. This clover is

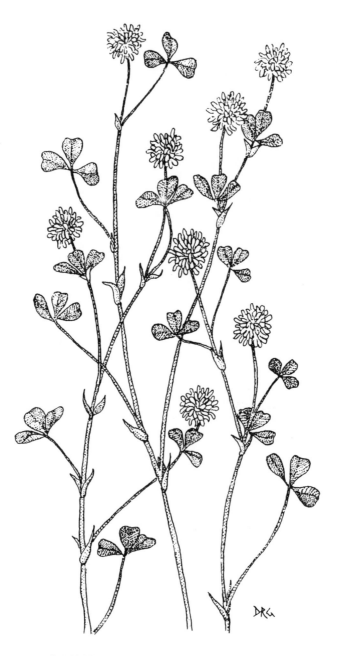

PERSIAN CLOVER
Trifolium resupinatum

reported to lodge easily.

USES: Hay, pasture, silage and green manure.

RANGE: Persian Clover grows well in the southern states
and in the coastal regions of the Pacific states. It
has produced well under irrigation in Arizona.

SOIL: This species is best adapted to low lying,
heavy, moist soils.

SEEDING: Sow at a rate of 4 to 6 lb.s per acre.
/h6, p4, p6, w4/

Trifolium subterranean SUB CLOVER, SUBTERRANEAN CLOVER

HABIT: A winter annual legume. The flowers of Sub
Clover are self fertilized. An unusual characteristic
of this plant is that the flowers burrow themselves
into the soil, like Peanuts', burying its own seed.
Sub Clover reseeds itself well. Livestock should be
removed at flowering to permit reproduction.

USES: Hay, pasture, cover crop and green manure.

RANGE: Sub Clover is native to the Mediterranean
region. It is adapted to climates with cool, moist
winters and dry summers, such as occur in California
and Oregon. Once established, the seedlings can
survive 10°F. Areas with 20 inches of annual rainfall
and below 3,300 feet in elevation are best for Sub
Clover.

        Australian Sub Clover selection and breeding
programs have produced a wide diversity of Sub Clover
strains. One of the major differences between strains
is the length of time required for the plant to produce
seed. The difference may be as great as 60 days between
the earliest and latest strains. These programs have
made it possible to find strains adapted over a wide
range of rainfall zones and soils.

SOIL: Generally, Sub Clover grows best in well
drained, loam soils. An important characteristic of
Sub Clover is that it grows well on hilly land. This
plant also is especially tolerant of acid soils. Sub
Clover has a medium lime requirement and responds well

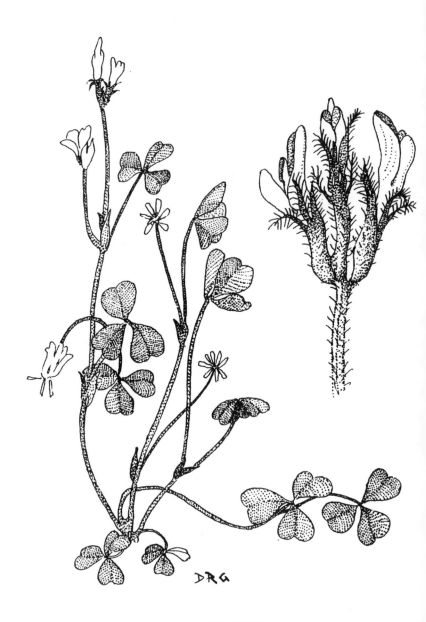

SUB CLOVER
Trifolium subterranean

to phosphate fertilization.
SEEDING: Sow in the fall. A recommended seeding rate is
  20 to 30 lb.s per acre at a depth of $\frac{1}{2}$ inch. The seed
  has a medium relative longevity.
  /a5, a9, b2, h6, m11, p4, p6, r1, w4/

SUB CLOVER STRAINS

Early	Geraldton, Dwalganup, Daliak
Mid season	Dinninup, Woogenellup, Mt. Barker
Late	Tallarock - 5 months to mature.
High water	Yarloop
High lime	Clare

Trifolium vesiculosum ARROWLEAF CLOVER
HABIT: A winter annual legume.
USES: Pasture and cover crop. Arrowleaf Clover is
  an important pasture plant on droughty, upland soils
  of the lower South where White Clover does not persist.
  It produces less, early, winter growth than Crimson
  Clover, but remains productive 6 to 8 weeks longer.
  A mixture of Arrowleaf and Crimson Clovers extends
  Clover production over a longer period of time.
RANGE: It grows in the Southern and Western states.
SOIL: Arrowleaf Clover will grow in any well drained
  soil, except soils with a high lime content. On soils
  with a pH of 7.5 or higher the plants become chlorotic
  or yellowish and make little growth.
  /h6, L2/

Trigonella foenum-graecum FENUGREEK
HABIT: A winter annual legume.
USES: Green manure and seed for seasoning. Fenugreek
  has been grown in California as an orchard green manure.
RANGE: The Southwestern states and California. Fenugreek grows best near the California seacoast. However,
  it has succeeded in all Citrus districts.
SOIL: This legume grows best in loam soils, yet it is
  not very exacting in its soil requirements. Fenugreek

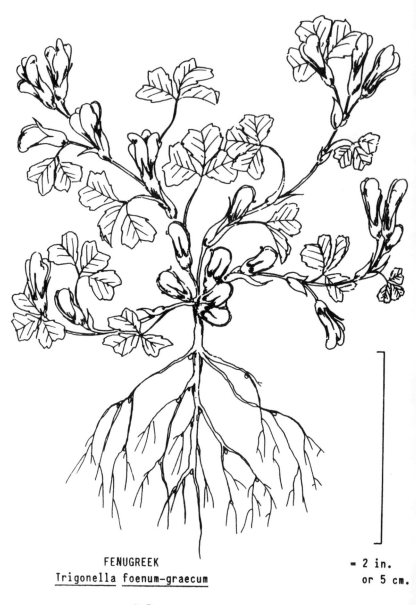

FENUGREEK
Trigonella foenum-graecum

K.B.

= 2 in.
or 5 cm.

WHEAT
Triticum aestivum

has a low lime requirement.
SEEDING: Sow in the fall. A recommended seeding rate is
  25 to 35 lb.s per acre at a depth of ½ inch. The seed
  remains viable for a relatively long time.
  /a5, h6, p4, p5/

Triticum aestivum WINTER WHEAT
HABIT: An annual cereal.
USES: Hay, grain, cover crop and green manure. As a
  hay, Wheat is inferior to Oats.
RANGE: Wheat originates from the Mediterranean region.
  In regions where autumn rains are favorable and winter
  is not severe, Winter Wheat can be grown. Winter Wheat
  is more winter hardy than Winter Barley, although Winter
  Wheat is less drought resistant. This cereal grows well
  at high elevations.
SOIL: Growth is best in loam soils. It has a low lime
  requirement.
SEEDING: Sow in the fall. A recommended seeding rate is
  100 lb.s per acre at a depth of 3/4 inches. The seed
  remains viable for a relatively long time.
  /a5, a7, h6, m10, s4/

CEREAL WINTER HARDINESS
most	Rye
	Wheat
	Barley
least	Oats

Vicia atropurpurea (V. benghalensis) PURPLE VETCH
HABIT: An annual legume. Purple Vetch is difficult to
  mow due to its viney growth.
USES: Forage, cover crop and green manure. It has
  been a successful green manure in California Citrus.
RANGE: Purple Vetch originates from Southern Europe.
  It is one of the least winter hardy Vetches. With
  fluctuating temperatures, it is injured at 10 to 15° F.
  Purple Vetch grows best in cool, moist weather. It

grows in all regions of the U.S.. This species grows
particularly well in California, Oregon and Washington.
In the milder parts of California, it is winter hardy,
while in Oregon and Washington, it occasionally winter
kills. Purple Vetch has been grown for seed in Western
Oregon and Washington and in Northwestern California.

SOIL: Best growth occurs on well drained loam or clay
loam soils. It has also grown fairly well on sandy
or gravelly soils. Purple Vetch has a low lime requirement.

SEEDING: In regions with mild winters, sow in the fall
for winter growth. In areas with severe winters, sow
in the spring for summer growth. A recommended seeding
rate is 50 to 60 lb.s per acre at a depth of 3/4 inches.
The seed remains viable for a relatively long time.
/a5, a9, b2, h3, h6, p4, p5/

VETCH WINTER HARDINESS

most	Hairy Vetch
	Hungarian Vetch
	Common Vetch
least	Purple Vetch

Vicia cracca BIRD VETCH, TUFTED VETCH

HABIT: A perennial legume. It has the same adaptations
as Hairy Vetch.

USES: Forage.

RANGE: A cool, moist climate.

SOIL: Bird Vetch will grow in any well drained soil,
including soils of low fertility.

SEEDING: A recommended seeding rate is 30 to 35 lb.s per
acre. The seed remains viable for a relatively long
time.
/a9, h6, p4/

Vicia dasycarpa WOOLLY POD VETCH (Lana, Auburn and
Oregon are cultivars)

HABIT: An annual legume. Woolly Pod Vetch matures

BIRD VETCH
Vicia cracca

AA | ×1

×4

earlier than Hairy Vetch and seeds abundantly.

USES: Forage, cover crop and green manure. Woolly
  Pod Vetch smothers out weed competition.

RANGE: Woolly Pod Vetch grows in all areas of the U.S.
  during cool, moist weather. This species has been
  grown for seed in California and Western Oregon. It is
  less winter hardy than Hairy Vetch, although Woolly Pod
  grows more prolifically in cool weather and is more
  heat tolerant. Where temperatures do not fluctuate
  violently or there is protection of snow, it will
  stand 0°F or lower.

SOIL: Woolly Pod Vetch grows well on any soil, even
  poor, sandy lands. It has a low lime requirement.

SEEDING: A recommended seeding rate is 50 to 60 lb.s
  per acre at a depth of 3/4 inches. The seed remains
  viable for a relatively long time.
  /a5, a9, b2, h3, h6, p4, p5/

Vicia ervilia BITTER VETCH, ERVIL, BLACK BITTER VETCH
         Bitter Vetch differs from the other Vetches in
  that it is more nearly upright in growth and does not
  have tendrils. It has never been considered superior
  to the other Vetches in common use. With fluctuating
  temperatures, it is injured at 10 to 15°F. Bitter
  Vetch has succeeded in California when fall sown.
  /h3, p4/

Vicia faba (V. faba var. minor, the small seeded variety)
  HORSE BEAN, FAVA BEAN, BROAD BEAN, WINDSOR BEAN, FIELD
  BEAN, BELL BEAN, TICK BEAN

HABIT: A winter annual legume.

USES: Silage, hay, cover crop, green manure, vegetable
  and grain. When grown with Corn and Sunflowers, Horse
  Beans make a good silage. It is not as valuable when
  grown for hay. As a green manure crop, this plant adds
  a large quantity of organic matter to the soil. The
  more succulent the vegetation, the more readily it
  decomposes. The greatest percentage of moisture is at

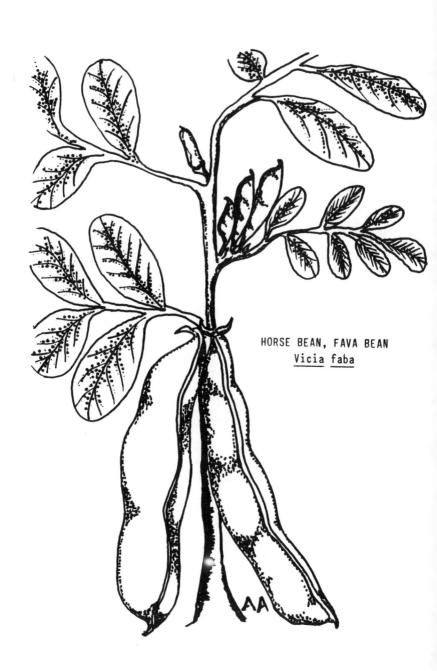

HORSE BEAN, FAVA BEAN
Vicia faba

blossom. Research has shown that six weeks of Horse
Bean growth fixes just as much nitrogen (100 lb.s per
acre) as other legumes do in a whole season.

RANGE: Horse Beans are adapted to a cool, temperate
  climate with an abundance of rainfall. It needs
  sufficient and uniform supplies of water and nutrients.
  This species grows best in the Maritime provinces of
  Canada and along the Pacific coast, especially the
  coastal strips of central California. Horse Beans do
  not endure heat. Seed production is curtailed by high
  temperatures.
         This plant is resistant to frost injury. The
  hardier varieties withstand temperatures as low as 15°F.
  In general, Horse Beans cannot be grown successfully
  where the temperature fluctuates rapidly.

SOIL: Best growth occurs on well drained, heavy silt
  or clay loams with large quantities of humus and
  calcium. Horse Beans will succeed on light soil with
  ample moisture and lime.

SEEDING: A recommended spacing is 30 inches between rows
  and 6 inches between plants. Sow 125 lb.s of seed per
  acre. Horse Beans may be grown in a mixture with
  Fenugreek, Vetch and Oats, drilling the Horse Beans
  two to three weeks earlier.
  /b2, h1, h3, k2, L4, p4, p5, r1, s3/

Vicia monanthos MONANTHA VETCH

HABIT: A winter annual legume. Monantha Vetch has
  finer stems than the other Vetches. It matures early
  and produces heavy yields of seed.

USES: Cover crop and green manure.

RANGE: Monantha Vetch survives the winters in the
  milder parts of Washington, Oregon and California.
  With fluctuating temperatures, it suffers injury at 10
  to 15°F.

SOIL: This species grows well on poor, sandy soil.
  /h3/

Vicia pannonica HUNGARIAN VETCH
HABIT: An annual legume.
USES: Forage and green manure.
RANGE: Hungarian Vetch is a native of Central Europe.
  It grows during the cool, moist season in all regions
  of the U.S., except Florida and the Gulf coast states.
  Hungarian Vetch is more winter hardy than Woolly Pod,
  but less so than Hairy. where temperatures do not
  fluctuate rapidly or there is protection of snow, it
  will withstand 0°F or lower. Without snow cover, it is
  safe only down to 10°F. Hungarian Vetch will not endure
  hot weather.
SOIL: Hungarian Vetch grows better in wet places or on
  poorly drained land than any other Vetch. Best growth
  occurs on clay loam or clay soils. Hungarian Vetch
  grows well on the white soils of Oregon where Common
  Vetch will not grow. This species has a low lime
  requirement.
SEEDING: Sow in fall or spring. A recommended seeding
  rate is 70 to 80 lb.s per acre at a depth of 3/4 inches.
  The seed remains viable for a relatively long time.
  /a5, a8, a9, h3, h6, p4, p5/

Vicia sativa COMMON VETCH (Willamette is a cultivar)
HABIT: An annual legume.
USES: Forage, hay and green manure.
RANGE: Common Vetch requires a cool growing season.
  It languishes in hot summer weather. This species
  grows in all areas of the U.S.. In California,
  Common Vetch grows best south of the Tehachapi mountains
  and in sheltered, isolated valleys.
          Common Vetch is less winter hardy than Hairy
  Vetch. However, in Oregon and Washington, it survives
  most winters. Without protection, none of the Common
  Vetch varieties withstand 0°F. This plant is useful
  where the winter temperature does not fall below 10°F.
SOIL: Common Vetch grows in almost any moist, well
  drained soil. Best growth occurs in loam or sandy

R.G.

COMMON VETCH
Vicia sativa

loam soils. It has a low lime requirement. On poor
soils, it is important to inoculate the seed.
SEEDING: In the Southern and Pacific coast states, sow
in the fall. In the Northern states and Canada, sow in
the spring. A recommended seeding rate is 70 to 80 lb.s
per acre at a depth of 3/4 inches. The seed has a
medium relative longevity.
/a5, a9, h3, h6, L1, p4, p5, w3/

Vicia sativa var. nigra (V. angustifolia) NARROWLEAF
 VETCH, AUGUSTA VETCH
HABIT: An annual legume.
USES: Pasture and green manure. This legume seldom
 succeeds under cultivation.
RANGE: Narrowleaf Vetch is naturalized in the Eastern
 U.S.. It is a weed in the Spring Wheat and Cotton
 belts. With fluctuating temperatures, Narrowleaf
 Vetch suffer injury at 10 to 15^{0}F.
SOIL: It succeeds only where there is fertile soil
 or plenty of organic matter.
 /h3/

Vicia villosa HAIRY VETCH, WINTER VETCH
HABIT: An annual legume which often acts as a biennial.
 The hairier varieties are more winter hardy. Vetches
 are rather shallow rooted.
USES: Forage, hay and green manure. Hairy Vetch is
 often grown with Wheat, Oats or Rye. These grasses act
 as nurse crops, protecting the young Vetch. As a
 green manure in mild winter climates, it is less
 desirable than Common or Purple Vetch because it makes
 less growth in the winter. Nevertheless, this Vetch is
 well known for fixing large quantities of nitrogen, as
 much as 100 lb.s per acre.
RANGE: Hairy Vetch will grow in all areas of the U.S.,
 during the cool, moist season. It is the most winter
 hardy of the cultivated Vetches. This is the only

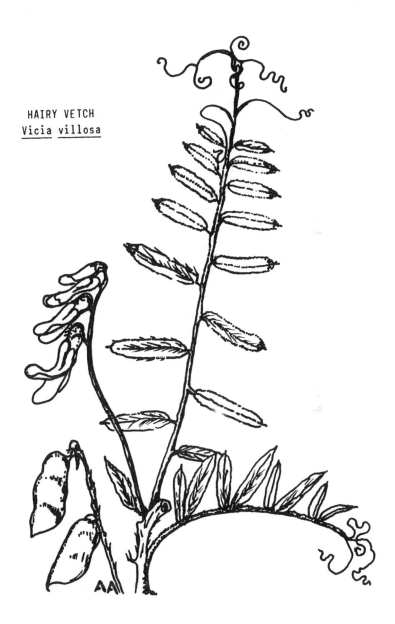

HAIRY VETCH
Vicia villosa

Vetch recommended for fall sowing in the North. In the
South, Hairy Vetch does not start growth, as other
legumes do, during the temporary warm spells. Consequently, it is not as likely to be killed by severe
frosts later. Hairy Vetch is grown for seed in
Western Oregon.

SOIL: Hairy Vetch will grow on any well drained soil,
  including soils of low fertility. It will grow on acid
  soils where Clover and Alfalfa will not. This plant
  also tolerates alkaline soils.

SEEDING: A recommended seeding rate is 40 to 60 lb.s
  per acre at a depth of 3/4 inches. The seed remains
  viable for a relatively long time. If allowed to set
  seed, Hairy Vetch will volunteer the next year.
  /a5, a7, a9, h3, h6, L2, m4, m7, p4, p5/

Vigna sinensis COW PEA
HABIT: A summer annual legume.
USES: Grain, hay, pasture, cover crop and green
  manure. There are many Cow Pea varieties. The more
  vigorous varieties can compete with grass weeds. Cow
  Peas serve the same purpose in the South as Red Clover
  does in the North. As a green manure, they are sometimes rolled down before being turned under.

RANGE: The Cow Pea is adapted to almost the same hot,
  moist climatic conditions as Corn, although it requires
  more heat. Cow Peas are slightly more drought resistant than Soybeans. They cannot be grown for grain
  as far north as Soybeans because they are more sensitive to frost.

SOIL: Cow Peas are adapted to a wide range of soil
  types, including low fertility soils. For optimum
  growth, they need good drainage, a sufficient supply
  of water and good aeration. Most of the moisture used
  by this plant comes from the surface foot of soil.
  Best growth occurs on sandy loam soil.

          The optimum pH for this crop is between 5.5
  and 6.5. Cow Peas have a low lime requirement. On

COW PEA
Vigna sinensis

soils too poor and too acid for Clover, Cow Peas grow
well. They are better than Clover or Alfalfa on thin
soils or soils poor in lime.

    For grain, a poor soil is better because a poor
soil produces little vegetation and ample seed, while
rich soil produces abundant vegetation and little seed.

    Cow Peas are tolerant of moderate shade. They
may be grown intermixed with Corn or in orchards. For
example, Cow Peas can be planted at 6½ feet square
spacings after the second or third cultivation of the
Corn intercrop. It is important to note that this
legume is subject to mildew in heavy shade.

SEEDING: Sow when the soil is thoroughly warmed and the
weather settled. When growing Cow Peas for the first
time in the North or West, inoculate the seed. A
recommended spacing for a grain crop is 3 feet between
rows with 2 to 3 inches between plants, at a rate of
30 to 40 lb.s per acre. For hay or green manure, the
seed can be drilled in rows 6 to 8 inches apart or
broadcast at a rate of 90 lb.s per acre. Plant the seed
at a depth of ½ to 1 inch, or deeper when sowing in
sandy soil. The seed remains viable for a relatively
short time.

/a5, a7, c1, h1, h6, m2, m6, p4, p5, s1, s3, s7, w3, w4/

Zea mays CORN, MAIZE
HABIT: A summer annual cereal.
USES: Grain, pasture, silage and green manure. When
  Corn is grown for silage or pasture, it is often grown
  with other plants, such as Soybeans, Cow Peas, Velvet-
  beans or Peanuts.
RANGE: Corn originates from northwestern South America
  and northern Central America. It requires a hot, moist
  climate. For grain, this plant needs 90 frost free
  days to mature. The minimum annual rainfall needed
  for growth is 20 to 25 inches.
SOIL: Corn is adapted to a wide range of soil types.
  This is a heavy feeding crop which grows best on fertile

soils that are well supplied with nitrogen and phosphorus. Corn has a low lime requirement.
SEEDING: Sow when the soil is warm. A recommended green
  manure seeding rate is 90 lb.s per acre at a depth of
  1 inch. The seed has a medium relative longevity.
  /a5, p4, s4, w2, w3/

In the Mouth
of the Future

The years passed. Hylas grew older, wiser and more
content. The five acres had been paid off early, so he
decided to cultivate ten more acres. His farming ability
had improved to such a degree that he was ready to expand.
Much of his success in farming, he attributed to green
manuring. Hylas never went one year without growing a
green manure crop. He always remembered to feed his
friends in the soil - particularly Eatmore's children,
grand children and great grand children. Eatmore had
passed away. Hylas missed their talks. Nevertheless,
he always had an earthworm to talk to. Eatmore's descen-
dents constantly badgered Hylas with questions about
Eatmore. Hylas may have gotten tired of the questions,
but he had such fond memories of Eatmore that he didn't
mind ruminating over them with the young earthworms. In
fact, he delighted in fascinating the earthworms with
Eatmore stories.

You might think 15 acres would be too much for one
rabbit to handle. It would have been, but Hylas now had
three young cottontails to help him. Yes, Hylas had
gotten married - not to Dahlia though, nor Violet, but
Lilac, a beautiful, long haired Angor rabbit who cooked
like a chef. Not long after that second growing season
had started, Dahlia found another lover. Hylas was
quickly lost from the picture. Dahlia became so enthralled with this new romance that she got married only a
few months later. As for Violet, by the middle of the
summer, she tired of the city. She was recaptured by
the spirit of Wanderlust. She headed for the tropics.

At first, Hylas was emotionally decimated by such

a loss of loves. He consoled himself by repeating things
like, "There are other fish in the sea." and immersing
himself in work. Work became such an all-encompassing
part of his life that Hylas didn't even come in contact
with the other fish in the sea. All that work paid off
though. Hylas became well known in the community for
his farming talent. He was a perennial figure on the
community council.

Within a few years, this rabbit became such a
famous seasonal farmer that he was hired to make winter
lecture tours, visiting seasonal farmers in the city.
These conferences were extremely beneficial to all the
farmers because it gave them the chance to exchange
ideas with other farmers whom they would never see during
the busy growing season. It was at one of these lectures
that Hylas met his wife, Lilac.

The community of seasonal farmers had blossomed too.
During the past five years, the seasonal farmers worked
to build a truely, magnificent community center. The
farmers slept in teepees on their land, while all their
other activities took place at the community center. It
was the hub of their lives. The center was not large by
any means. It was just the right size for the twenty
five families of rabbits and chickens who built it. The
center met almost all of the needs of the community. It
had a kitchen and dining area, general store with a bank
and post office, a gymn - which was really a large,
multipurpose room that could be divided up - with a sauna
and a hot tub, and a health center. At the health center,
a full time health practitioner cured minor illnesses,
cared for minor emergencies and taught classes. The
health of the community was largely left in the hands of
the health practitioner. She took a preventative approach
toward medicine.

A travelling doctor made frequent visits to the
community. She dealt with problems which the health
practitioner could not handle alone and generally gave
her advice. For the rare occurence of severe illnesses

or emergencies, the patient was sent to a hospital in the
city, or a travelling doctor visited patients who couldn't
travel.

"Health, Beauty and Permanence, eh?" Hylas and a
reporter had just stepped out of the front door of the
community center. The reporter, whose name was Grossmund,
was admiring the plaque which hung over the door. "I
understand why you have health and beauty hanging above
your door, Hylas. After all, most folks these days
think these two qualities are synonymous. Maybe you'd
like to explain what permanence means to your community."

"Well Grossmund, to us permanence has a two-fold
meaning. First, permanence means durable. That is, we
strive to build things so well, that they will last a
lifetime, or even longer in most cases. We build and
maintain things to last as long as possible. Second,
permanence means recyclable. This is the meaning of the
word which takes the earth into consideration as a living
organism. Materials we use, fit into the living system of
the organism, earth. The living earth never creates anything which cannot be broken down and recreated into
another life form. You might think that this is not
permanent at all, but change or dynamism. For us, this
means promoting permanent change, so our community lives
dynamically within the design of the global organism,
earth, rather than outside of it. We feel like cells
of a large organism. All the cells are related. Our
community strives always to be productive, harmonious
cells of the one body. All of the materials we use are
recyclable in some way. In fact, if you asked one of
our children what trash or garbage meant, they wouldn't
have the vaguest notion of what these words stand for.
Even our own fecal waste is transformed into usable
fertilizer and biogas."

"That leads me to my next question, Hylas. What
about the children? What do they spend their time doing?"

"The children are kept out of, or should I say in,
mischief during the winter months (November through March)

by attending school full time. They go to mostly academic and cultural classes. Then, during the growing season, they go to practical schools for half days. The children choose whatever subjects they wish. Some of the practical courses offered are: farming, of course – which include both plant and animal husbandry, also cooking, carpentry, health and many other fields. The summer lessons are really closer to apprenticeships than anything else.

The bunnies and chicks do put in their fair share on the work, I might add. They are a big help too. The three big chores which the children work on are weeding, harvesting and processing food. A full day of work is never expected of a child. However, as they grow older, they can work longer and accept more responsibility. For example, the young adults who, at a younger adolescent age, cut fruit for drying and canning, now assist in sterilizing equipment and mixing the ingredients with the head cook. Processing food is an integral part of our community of seasonal farmers, as it is to all communities. We must meet high standards for the export and local markets."

"Now wait Hylas. Let's take each of these markets in turn. Tell me about the export market of your produce."

"Grossmund, as you probably know, the distribution and marketing of food around the planet has been vastly improved in the past five years. We are now concentrating more of our effort in growing protein crops for export – crops like nuts, grains and dairy products – protein which can be transported long distances without spoiling. Why, food is one of our most valuable resources, along with soil, water and people. We are beginning to utilize these resources more efficiently on a global level."

"What about energy?" asked Grossmund.

"I know at one time there was a big stir about energy," began Hylas. The energy situation has come into balance once again, mainly because we have abandoned

fossil fuels for renewable fuels which we can grow or in
some way make ourselves. Also, we are much more efficient
energy users today. We don't waste. Take our community
for example, our energy comes from solar collectors,
biogas and alcohol. Alcohol production is the most
fascinating to me. We can take any cellulose source –
mainly we get it from swamp vegetation and from harvesting forest leaves – and make butanol out of it. Energy
was once a lion of a problem, but it is now a kitten.

To continue with my marketing explanation: Our
export and local marketing are really handled in about the
same manner. As each of our crops is harvested, a review
of our produce appears on a computer comodities listing.
We do not grow food with the aim of producing just quantity, but also quality. You see, all of our crops are
tested for nutritional quality – contents of carbohydrates, proteins, vitamins and minerals. All of this
information appears on the computer listing. The buyers,
which include both large and small organizations, such
as: urban buying clubs, coops, super markets, and export
merchants, look at the listing. Sometimes, they visit
the farms. Then a computer auction is held. The food
goes to the highest bidder. Today's marketing system is
very equitable. Farmers are rewarded justly for their
work and skill in producing quality along with quantity."

"Everything you've said so far sounds positive and
productive," applauded Grossmund. "Now, what is wrong
with your community? I mean, what are some of your most
outstanding difficulties?"

"Well, do you have a few spare hours? I love to
complain." Hylas quipped. As he said this, a gust of
wind blew the papers out of Grossmund's hands. They
glided.

Common Names — Latin Names

Adzuki bean	Phaseolus angularis
Alfalfa – Common	Medicago sativa
Alsike clover	Trifolium hybridum
Alyceclover	Alysicarpus vaginalis
Annual ryegrass	Lolium multiflorum
Annual yellow clover	Melilotus indica
Arrowleaf clover	Trifolium vesiculosum
Augusta vetch	Vicia sativa var. nigra
Austrian brome	Bromus inermis
Austrian winter pea	Pisum arvense
Avare	Dolichos lablab
Ball clover	Trifolium nigrescens
Barley	Hordeum vulgare
Bean	
Adzuki	Phaseolus angularis
Bell	Vicia faba
Broad	V. faba
Butter	Phaseolus limensis
Common	P. vulgaris
Dolichos	Dolichos lablab
Fava	Vicia faba
Field	V. faba
Field	Phaseolus vulgaris
Haricot	P. vulgaris
Horse	Vicia faba
Hyacinth	Dolichos lablab
Jack	Canavalia ensiformis
Kidney	Phaseolus vulgaris
Lima	P. limensis
Mat	P. aconitifolius

Bean
    Moth   Phaseolus aconitifolius
    Mung   P. aureus
    Pink   P. vulgaris
    Pinto   P. vulgaris
    Red Mexican   P. vulgaris
    Soy   Glycine max
    Sword   Canavalia gladiata
    Tepary   Phaseolus acutifolius
    Texas   P. acutifolius
    Tick   Vicia faba
    Urd   Phaseolus mungo var. radiatus
    Velvet·   Stizolobium deeringianum
    Windsor   Vicia faba
Beggarweed   Desmodium purpureum
Bell bean   Vicia faba
Bermudagrass   Cynodon dactylon
Berseem clover   Trifolium alexandrinum
Bighop clover   Trifolium procumbens
Bird vetch   Vicia cracca
Bird'sfoot trefoil
    Big   Lotus pendunculatus
    Broadleaf   L. corniculatus
    Greater   L. major
    Marsh   L. pendunculatus
    Narrowleaf   L. tenuis
Bitter clover   Melilotus indica
Bitter melilot   M. indica
Bitter vetch   Vicia ervilia
Black gram   Phaseolus mungo var. radiatus
Black medic   Medicago lupulina
Bluegrass
    Canada   Poa compressa
    Kentucky   P. pratensis
    Virginia   P. compressa
Bokhara sweet clover Melilotus alba
Borage   Borago officinalis
Broad bean   Vicia faba

Bromegrass
    Austrian   Bromus inermis
    Blando   B. mollis
    Bromar mountain   B. marginatus
    California   B. carinatus
    Cucamonga   B. carinatus
    Field   B. arvensis
    Hungarian   B. inermis
    Mountain   B. marginatus
    Prairie   B. catharticus
    Russian   B. inermis
    Smooth   B. inermis
Broom corn millet   Panicum miliaceum
Buckwheat   Fagopyrum esculentum
Bur clover   Medicago hispida
Button clover   M. orbicularis
Button medic   M. orbicularis
Caley pea   Lathyrus hirsutus
California bromegrass   Bromus carinatus
Canada bluegrass   Poa compressa
Canadian field pea   Pisum arvense
Cereal rye   Secale cereale
Cheatgrass   Bromus tectorum
Chess   B. tectorum
Chick pea   Cicer arientinum
Cicer milkvetch   Astragalus cicer
Clover
    Alsike   Trifolium hybridum
    Arrowleaf   T. vesiculosum
    Ball   T. nigrescens
    Berseem   T. alexandrinum
    Bighop   T. procumbens
    Bitter   Melilotus indica
    Bur   Medicago hispida
    Button   M. orbicularis
    California   M. hispida
    Crimson   Trifolium incarnatum
    Eygptian   T. alexandrinum

Clover
- Fieldhop — Trifolium agrarium
- Hungarian — T. pannonicum
- Japan — Lespedeza striata
- Ladino — Trifolium repens var. latum
- Persian — T. resupinatum
- Red — T. pratense
- Rose — T. hirtum
- Smallhop — T. dubium
- Sour — Melilotus indica
- Spotted bur — Medicago arabica
- Strawberry — Trifolium fragiferum
- Sub — T. subterranean
- Subterranean — T. subterranean
- Tick — Desmodium purpureum
- Toothed bur — Medicago hispida
- White — Trifolium repens
- Wood's — Dalea alopecuroides

Common alfalfa   Medicago sativa
Common bean   Phaseolus vulgaris
Common Lespedeza   Lespedeza striata
Common Sesbania   Sesbania macrocarpa
Common sweet clover   Melilotus officinalis
Common vetch   Vicia sativa
Corn   Zea mays
Cow Pea   Vigna sinensis
Crested wheatgrass   Agropyron desertorum
Crotolaria   Crotolaria spectabilis
Crown vetch   Coronilla varia
Cruickshank's lupine Lupinus cruickshankii
Cucamonga brome   Bromus carinatus
Dalea   Dalea alopecuroides
Deering velvetbean   Stizolobium deeringianum
Dolichos bean   Dolichos lablab
Eygptian clover   Trifolium alexandrinum
Eygptian lupine   Lupinus termis
English ryegrass   Lolium perenne
Ervil   Vicia ervilia

Esparsette	Onobrychis sativa
Fava bean	Vicia faba
Fenugreek	Trigonella foenum-graecum
Fescue - Tall	Festuca arundinacea
Field bean	Phaseolus vulgaris
Field bean	Vicia faba
Field bromegrass	Bromus arvensis
Field pea	Pisum arvense
Fieldhop clover	Trifolium agrarium
Flat pea	Lathyrus sylvestris
Flat-podded vetchling	L. cicera
Flatpod peavine	L. cicera
Florida beggarweed	Desmodium purpureum
Florida velvetbean	Stizolobium deeringianum
Foxtail	Dalea alopecuroides
Foxtail millet	Setaria italica
Fragrant lupine	Lupinus luteus
Frijolita de Cuba	Phaseolus limensis
Garbonzo	Cicer arientinum
Genge	Astragalus sinicus
Giant spurry	Spergula maxima
Golden gram	Phaseolus aureus
Gram pea	Cicer arientinum
Grass pea	Lathyrus sativus
Greater bird'sfoot trefoil	Lotus major
Green gram	Phaseolus mungo
Groundnut	Arachis hypogaea
Guar	Cyamopsis tetragonoloba
Hairy indigo	Indigofera hirsuta
Hairy lupine	Lupinus hirsutus
Hairy vetch	Vicia villosa
Haricot bean	Phaseolus vulgaris
Hemp Sesbania	Sesbania macrocarpa
Hog millet	Panicum miliaceum
Horse bean	Vicia faba
Horse gram	Dolichos bifloris
Hungarian brome	Bromus inermis
Hungarian clover	Trifolium nigrescens
Hungarian vetch	Vicia pannonica

Hyacinth bean	Dolichos lablab
Indigo - Hairy	Indigofera hirsuta
Italian millet	Setaria italica
Italian ryegrass	Lolium multiflorum
Jack bean	Canavalia ensiformis
Japan clover	Lespedeza striata
Japanese Lespedeza	L. striata
Kale	Brassica oleracea
Kentucky bluegrass	Poa pratensis
Kidney bean	Phaseolus vulgaris
Kidney vetch	Anthyllis vulneraria
King island melilot	Melilotus indica
Korean Lespedeza	Lespedeza stipulacea
Kudzu	Pueraria hirsuta
Lablab	Dolichos lablab
Ladino clover	Trifolium repens var. latum
Lentil	Lens esculenta
Lespedeza	
Common	Lespedeza striata
Japanese	L. striata
Korean	L. stipulacea
Sericea	L. cuneata
Lupine	
Cruickshank's	Lupinus cruickshankii
Eygptian	L. termis
Fragrant	L. luteus
Hairy	L. hirsutus
Large blue	L. pilosus var. caeruleus
Large white	L. albus
Narrow-leaved	L. angustifolius
Perennial	L. perennis
Pink	L. pilosus var. roseus
Small blue	L. angustifolius var. caeruleus
Small white	L. angustifolius var. diploleuca
Succulent	L. affinis
Sundial	L. perennis

Lupine
    Yellow   Lupinus luteus
Maize   Zea mays
Marsh bird'sfoot trefoil   Lotus pendunculatus
Mat bean   Phaseolus aconitifolius
Medic
    Black   Medicago lupulina
    Button   M. orbicularis
    Spotted   M. arabica
    Toothed   M. hispida
Milkvetch
    Cicer   Astragalus cicer
    Sickle   A. falcatus
    Sicklepod   A. falcatus
Millet   Pennisetum glaucum
Millet
    Broomcorn   Panicum miliaceum
    Foxtail   Setaria italica
    Hog   Panicum miliaceum
    Italian   Setaria italica
    Pearl   Pennisetum glaucum
    Proso   Panicum miliaceum
    Texas   P. texanum
Monantha vetch   Vicia monanthos
Moth bean   Phaseolus aconitifolius
Mountain bromegrass   Bromus marginatus
Mung bean   Phaseolus aureus
Narrowleaf bird'sfoot
    trefoil   Lotus tenuis
Narrowleaf vetch   Vicia sativa var. nigra
Narrow-leaved lupine   Lupinus angustifolius
Oats   Avena sativa
Ochrus   see Lathyrus cicera
Official melilot   Melilotus officinalis
Orchardgrass   Dactylis glomerata
Pea
    Austrian winter   Pisum arvense
    Caley   Lathyrus hirsutus

Pea
Canadian field	Pisum arvense
Chick	Cicer arientinum
Cow	Vigna sinensis
Field	Pisum arvense
Flat	Lathyrus sylvestris
Gram	Cicer arientinum
Grass	Lathyrus sativus
Pigeon	Cajanus cajan
Rough	Lathyrus hirsutus
Singletary	L. hirsutus
Southern winter	L. hirsutus
Tangier	L. tingitanus
Wagner	L. sylvestris

Peanut	Arachis hypogaea
Pearl millet	Pennisetum glaucum
Perennial lupine	Lupinus perennis
Perennial ryegrass	Lolium perenne
Persian clover	Trifolium resupinatum
Pigeon pea	Cajanus cajan
Pink lupine	Lupinus pilosus var. roseus
Prairie bromegrass	Bromus catharticus
Proso millet	Panicum miliaceum
Purple vetch	Vicia atropurpurea
Rape	Brassica napus
Red clover	Trifolium pratense
Redtop	Agrostis alba
Rescuegrass	Bromus catharticus
Rose clover	Trifolium hirtum
Rough pea	Lathyrus hirsutus
Russian bromegrass	Bromus inermis
Rye	Secale cereale

Ryegrass
English	Lolium perenne
Italian	L. multiflorum
Perennial	L. perenne
Swiss	L. rigidum
Wimmera	L. rigidum

Sainfoin	Onobrychis sativa
Scotch Kale	Brassica oleracea
Sericea Lespedeza	Lespedeza cuneata
Serradella	Ornithopus sativus
Sesbania	Sesbania macrocarpa
Showy Crotolaria	Crotolaria spectabilis
Sickle-milkvetch	Astragalus falcatus
Sicklepod milkvetch	A. falcatus
Silverleaf	Desmodium uncinatum
Singletary pea	Lathyrus hirsutus
Small blue lupine	Lupinus angustifolius var. caeruleus
Small white lupine	L. angustifolius var. diplo-leuca
Smallhop clover	Trifolium dubium
Smooth bromegrass	Bromus inermis
Soft chess	Bromus mollis
Sorghum	Sorghum bicolor
Sour clover	Melilotus indica
Southern winter pea	Pisum arvense
Soybean	Glycine max
Spotted burclover	Medicago arabica
Spotted medic	M. arabica
Spurry	Spergula arvensis
Spurry - Giant	S. maxima
Strawberry clover	Trifolium fragiferum
Sub clover	T. subterranean
Subterranean clover	T. subterranean
Succulent lupine	Lupinus affinis
Sudangrass	Sorghum bicolor var. sudanense
Sundial lupine	Lupinus perennis
Sunflower	Helianthus annuus
Sunn hemp	Crotolaria juncea
Sweet clover	
Bokhara	Melilotus alba
Common	M. officinalis
White	M. alba

Sweet clover	
Yellow	Melilotus officinalis
Swiss ryegrass	Lolium rigidum
Sword bean	Canavalia gladiata
Tall fescue	Festuca arundinacea
Tangier pea	Lathyrus tingitanus
Tepary bean	Phaseolus acutifolius
Texas bean	P. acutifolius
Texas millet	Panicum texanum
Tick bean	Vicia faba
Tickclover	Desmodium purpureum
Timothy	Phleum pratense
Toothed bur clover	Medicago hispida
Toothed medic	M. hispida
Tufted vetch	Vicia cracca
Turnip	Brassica rapa
Urd bean	Phaseolus mungo var. radiatus
Velvetbean	Stizolobium deeringianum
Vetch	
Augusta	Vicia sativa var. nigra
Bird	V. cracca
Black bitter	V. ervilia
Bitter	V. ervilia
Common	V. sativa
Crown	Coronilla varia
Hairy	Vicia villosa
Hungarian	V. pannonica
Kidney	Anthyllis vulneraria
Monantha	Vicia monanthos
Narrowleaf	V. sativa var. nigra
Purple	V. atropurpurea
Tufted	V. cracca
Winter	V. villosa
Woolly pod	V. dasycarpa
Virginia bluegrass	Poa compressa
Wagner pea	Lathyrus sylvestris
Wheatgrass – Crested	Agropyron desertorum
White clover	Trifolium repens

White sweet clover	Melilotus alba
Wimmera ryegrass	Lolium rigidum
Windsor bean	Vicia faba
Winter rape	Brassica napus
Winter vetch	Vicia villosa
Winter wheat	Triticum aestivum
Wiregrass	Poa compressa
Wood's clover	Dalea alopecuroides
Woolly pod vetch	Vicia dasycarpa
Yellow lupine	Lupinus luteus
Yellow sweet clover	Melilotus officinalis
Yellow trefoil	Medicago lupulina

✃ Reference Section ✄

a1.   Anon. 1978. THE WINGED BEAN. The California and
      Northwestern Organic Journal. April-May pp. 19-23
a2.   Anon. 1977. BUCKWHEAT. The California and
      Northwestern Organic Journal. Aug.-Sept. pp. 21
      & 39
a3.   Army, T.J. and J.C. Hide 1959. GREEN MANURE IS
      NOT BENEFICIAL TO DRYLAND. Crops and Soils.
      11(4): 23
a4.   Army, T.J. and J.C. Hide 1959. EFFECTS OF GREEN
      MANURE CROPS ON DRYLAND WHEAT PRODUCTION IN THE
      GREAT PLAINS AREA OF MONTANA. Agronomy Jour.
      51: 196-198
a5.   Alther, Richard and Richard O. Raymond 1974.
      IMPROVING GARDEN SOIL WITH GREEN MANURE. Charlotte,
      Vt.: Garden Way Pub. Co.
a6.   Alexander, Martin 1977. INTRODUCTION TO SOIL
      MICROBIOLOGY. N.Y.: Wiley Press p. 307
a7.   Ahlgren, Gilbert H. 1956. FORAGE CROPS. N.Y.,
      Toronto, London: McGraw-Hill Book Co. Inc.
a8.   Anon. 1977. COVER CROP GARDENING. SOIL ENRICH-
      MENT WITH GREEN MANURES. Charlotte, Vt.: Garden
      Way Pub. Co.
a9.   Anon. 1979. COMMON NORTHWEST LEGUMES. Tilth
      5(2): 7
b1.   Bailey, L.H. 1907. CLOVER. Cyclopedia of
      American Agriculture. pp. 232-239
b2.   Burroughs, Kate 1979. CALCIUM, COMPOST AND COVER
      CROPS, Rx FOR NORTH COAST SOILS. PART 2. Redwood
      Rancher. Holiday Issue. pp. 34-37

b3.   Brady, Nyle C. 1974. THE NATURE AND PROPERTIES OF
      SOIL. N.Y.: MacMillan Pub. Co.

b4.   Brinton, William F. Jr. 1981. SOIL TESTING:
      PROBING LIEBIGS LEGACY. Bio-Dynamics 139: 3-16

c1.   Coleman, Leslie C., B. Narasimha Iyengar and N.
      Sampatiengar 1912. GREEN MANURING IN MYSORE.
      Mysore Dept. of Ag. India. General Series Bull.
      #1

c2.   Cox, Jeff 1976. AZOTOBACTOR: THE SOIL BACTERIA
      THAT CAN INCREASE YOUR YIELDS. Organic Gardening
      and Farming. April

c3.   Cantisano, Robert   personal communication.

c4.   chesney, charles   personal communication.

d1.   Davy, J. Burt 1899. LUPINES FOR GREEN MANURING.
      Univ. of Ca. Ag. Exp. Sta. Bull. 124

e1.   Edwards, C.A. Oct. 1980. INTERACTIONS BETWEEN
      AGRICULTURAL PRACTICES AND EARTHWORMS. Proceedings
      of the 7th Int. Colloquim of Soil Zoology. pp. 3-
      12

f1.   Fred, E.B. 1916. RELATION OF GREEN MANURES TO
      THE FAILURE OF CERTAIN SEEDLINGS. U.S.D.A. Jour.
      of Ag. Res. 5 #25 pp. 1161-1176

f2.   Fukuoka, M. 1978. THE ONE STRAW REVOLUTION: AN
      INTRODUCTION TO NATURAL FARMING. Emmaus, Pa.:
      Rodale Press

f3.   Foth, Henry 1972. FUNDAMENTALS OF SOIL SCIENCE.
      N.Y.: John Wiley & Sons

g1.   Graves, Walter L., Burgess L. Kay and Tom Ham
      1980. ROSE CLOVER CONTROLS EROSION IN SOUTHERN
      CALIFORNIA. Cal. Ag. 34(4): 4-5

g2.   Gill, K.S., T.S. Sandhu and H.S. Sekhon 1978.
      GREEN MANURE PADDY FIELDS FOR HIGHER PRODUCTION.
      Indian Farming. 28(3): 19 & 30

g3.   Grossman, Joel 1980. A LOW SODIUM DIET FOR
      IRRIGATED SOILS. The New Farm. 2(7): 36-43·

g4.   Guenzi, W.D. (ed.) 1974. PESTICIDES IN SOIL AND
      WATER. S.S.S.A. Inc.

g5.   Golueke, Clarence G. 1972. COMPOSTING. A STUDY
      OF THE PROCESS AND ITS PRINCIPLES. Emmaus, Pa.:
      Rodale Press
h1.   Hardenburg, E.V. 1927. BEAN CULTURE. N.Y.: The
      MacMillan Co.
h2.   Hendry, G.W. 1918. BEAN CULTURE IN CALIFORNIA.
      Univ. of Ca. Ag. Exp. Sta. Bull. #294
h3.   Henson, P.R. and H.A. Schotch 1968. VETCH CULTURE
      AND USES. U.S.D.A. Farmers' Bull. 1740
h4.   Hill, H.H. 1926. DECOMPOSITION OF ORGANIC MATTER
      IN SOIL. U.S.D.A. Jour. of Ag. Res. 33 #1: 77-99
h5.   Hunt, Thomas F. 1907. THE FORAGE AND FIBER CROPS
      IN AMERICA. N.Y.: Orange Judd Co.
h6.   Heath, Maurice E., Darrel S. Metcalfe and Robert F.
      Barnes 1973. FORAGES, THE SCIENCE OF GRASSLAND
      AGRICULTURE. Ames, Iowa: The Iowa State Univ.
      Press
h7.   Holland, A.A. and J.E. Street 1968. PELLET INOCU-
      LATION OF LEGUME SEED. Univ. of Ca. Ag. Ext. Ser.
      AXT-280
h8.   Holland, A.A., J.E. Street and W.A. Williams 1969.
      RANGE-LEGUME INOCULATION AND NITROGEN FIXATION BY
      ROOT NODULE BACTERIA. Univ. of Ca. Ag. Exp. Sta.
      Bull. 842
j1.   Jensen, William A. and Frank B. Salisbury 1972.
      BOTANY: AN ECOLOGICAL APPROACH. Belmont, Ca.:
      Wadsworth Pub. Co.
k1.   Kennedy, P.B. 1922. LEGUMINOUS PLANTS AS ORGANIC
      FERTILIZERS IN CALIFORNIA AGRICULTURE. Univ. of Ca.
      Ag. Exp. Sta. Cir. #255
k2.   Kennedy, P.B. 1923. THE SMALL-SEEDED HORSE BEAN
      (Vicia faba var. minor). Univ. of Ca. Ag. Exp.
      Sta. Cir. #257
k3.   Kennedy, P.B. 1925. THE TANGIER PEA (Lathyrus
      tingitanus). Univ. of Ca. Ag. Exp. Sta. Cir. #290
L1.   Lipman, Chas. B. 1913. GREEN MANURING IN CALIFOR-
      NIA. Univ. of Ca. Ag. Exp. Sta. Cir. #110

L2. Logsdon, Gene 1980. FALL PLANTINGS THAT FEED YOUR GARDEN. Org. Gar. 27(9): 128-141

L3. Logsdon, Gene 1977. SMALL-SCALE GRAIN RAISING. Emmaus, Pa.: Rodale Press

L4. Logsdon, Gene 1981. SPECIALTY LEGUMES MAINTAIN SOIL FERTILITY (WITHOUT SHUTTING OFF CASH FLOW). The New Farm. 3(1): 27-29

L5. Logsdon, Gene 1980. FITTING THE RIGHT LEGUME TO YOUR CROP ROTATION SYSTEM. The New Farm. 2(7): 44-47, 50-57

L6. Logsdon, Gene 1981. THE CLOVEREST WAY TO FARM YET. The New Farm. 3(4): 24-25

m1. Madson, B.A. 1916. A COMPARISON OF ANNUAL CROP-PING, BIENNIAL CROPPING AND GREEN MANURES ON THE YIELD OF WHEAT. Univ. of Ca. Ag. Exp. Sta. Bull. #270

m2. McKee, Roland 1935. SUMMER CROPS FOR GREEN MANURE AND SOIL IMPROVEMENT. U.S.D.A. Farmers' Bull. #1750

m3. McKee, Roland and Walter V. Kell 1947. COVER CROPS FOR SOIL CONSERVATION. U.S.D.A. Farmers' Bull. #1758

m4. McKee, Roland 1943. LEGUME COVER CROPS IN THE SOUTH. U.S.D.A. Ag. War Information #67

m5. McKee, Roland 1949. BUR CLOVER CULTIVATION AND UTILIZATION. U.S.D.A. Farmers' Bull. #1741

m6. McKee, Roland 1942. SUMMER CROPS FOR GREEN MANURE AND SOIL IMPROVEMENT. U.S.D.A. Bull. #1758 (Ref. m2 revised)

m7. McWhorter, O.T., E.R. Jackman and Arthur King 1945. ORCHARD SOIL COVERS. Or. State Col. Ext. Bull. #650

m8. Mertz, W.M. 1915. <u>Melilotus</u> <u>indica</u> AS A GREEN MANURE CROP IN SOUTHERN CALIFORNIA. Univ. of Ca. Ag. Exp. Sta. Cir. #136

m9. Mertz, W.M. 1918. GREEN MANURE CROPS IN SOUTHERN CALIFORNIA. Univ. of Ca. Ag. Exp. Sta. Bull. #292

m10. Morrison, K.J., Frank J. Viets Jr. and C.E. Nelson
      1954. GREEN MANURE AND COVER CROPS FOR IRRIGATED
      LAND. Wa. State Ext. Bull. #489

m11. Murphy, A.H., M.B. Jones, J.W. Clawson and J.E.
      Street 1973. MANAGEMENT OF CLOVERS ON CALIFORNIA
      ANNUAL GRASSLANDS. Univ. of Ca. Ag. Exp. Sta. Ext.
      Ser. Cir. #564

m12. Muzick, Mark 1980. NITROGEN FIXING TREES AND
      SHRUBS FOR THE PACIFIC NORTHWEST. Arlington, Wa.:
      Tilth

m13. Mengel, Konrad and Ernest A. Kirby 1978. PRINCI-
      PLES OF PLANT NUTRITION. Berne, Switzerland: Int.
      Potash Inst.

m14. McLeod, Edwin J. and Sean L. Swezey 1979. SURVEY
      OF WEED PROBLEMS AND MANAGEMENT TECHNOLOGIES IN
      ORGANIC AGRICULTURE. P.O. Box 475, Graton, Ca.:
      Organic Agriculture Research Inst.

m15. McLeod, Edwin J. 1980. AN ANNOTATED BIBLIOGRAPHY
      OF NON CHEMICAL WEED MANAGEMENT. Graton, Ca.:
      Organic Agriculture Research Inst.

m16. McSweeney, Kevin   personal communication

m17. Minnich, Jerry, Marjorie Hunt and the Editors of
      Organic Gardening Magazine. 1979. THE RODALE
      GUIDE TO COMPOSTING. Emmaus, Pa.: Rodale Press

n1.   National Academy of Sciences, Nat. Res. Council.
      1979. TROPICAL LEGUMES: RESOURCES FOR THE FUTURE.
      Washington D.C.: National Academy of Sciences

o1.   Organic Gardening and Farming Staff. 1973.
      ORGANIC FERTILIZERS: WHICH ONES AND HOW TO USE
      THEM. Emmaus, Pa.: Rodale Press

p1.   Peterson, M.L., J.E. Street and V.P. Osterli 1962.
      SALINA STRAWBERRY CLOVER. Univ. of Ca. Ag. Exp.
      Sta. Ext. Ser. Leaflet #2146 or 146 old system.

p2.   Pieters, A.J. and Roland McKee 1938. THE USE
      OF COVER AND GREEN MANURE CROPS. Soils and Men.
      U.S.D.A. Yearbook of Ag. 1938.

p3.   Piper, C.V. 1934. IMPORTANT CULTIVATED GRASSES.
      U.S.D.A. Farmers' Bull. #1254

p4.   Piper, Charles V. 1924. FORAGE PLANTS AND THEIR
      CULTURE. N.Y.: MacMillan Co.

p5.   Pieters, Adrian J. 1927. GREEN MANURING, PRINCI-
      PLES AND PRACTICE. N.Y.: John Wiley and Sons Inc.

p6.   Phillips Petroleum Co. 1960. INTRODUCED GRASSES
      AND LEGUMES. Section 5 & 6 of a series of pasture
      and range plants. Bartlesville, Oklahoma: Phillips
      Petroleum Co.

r1.   Robinson, D.H. 1937. LEGUMINOUS FORAGE PLANTS.
      London: Edward Arnold Co.

r2.   Ritchie, Jim 1980. THE BIG WINTER COVER-UP.
      The New Farm 2(6): 94-96

r3.   Rogers, T.H. and J.E. Giddens 1957. GREEN MANURE
      AND COVER CROPS. U.S.D.A. Yearbook of Ag. pp. 252-
      257

r4.   Rodale, J.I. and Staff 1976. 13th Printing.
      ENCYCLOPEDIA OF ORGANIC GARDENING. Emmaus, Pa.:
      Rodale Books Inc.

r5.   Ray, Peter Martin 1972. THE LIVING PLANT.
      N.Y.: Holt, Rinehart and Winston Inc.

r6.   Ritchie, James D. 1981. ROTATIONS SOLVE PROBLEMS
      YOU CAN'T CURE OUT OF A SACK. The New Farm 3(6):
      42-45

s1.   Sallee, William R. and Francis L. Smith 1969.
      COMMERCIAL BLACKEYE BEAN PRODUCTION IN CALIFORNIA.
      Univ. of Ca. Ag. Exp. Sta. Ext. Ser. Leaflet 2376
      or old system Cir. 549

s2.   Schade, Louis 1861. CHARACTER, CULTIVATION AND
      USE OF THE LUPINE. Report of the Dept. of Ag.
      pp. 370-373

s3.   Schafer, Paul 1967. THE MANURING OF PULSES.
      Green Bull. #20

s4.   Schery, Robert W. 1952. PLANTS FOR MAN. Englewood
      Cliffs, N.J.: Prentice-Hall Inc.

s5.   Shaw, Thomas 1906. PROFITABLE HAY MAKING. In
      Farm Science. Chicago, Ill.: Int. Harvester Co.
      of America

s6.   Singh, N.T., S.M. Verma and A.S. Josan 1975.
      DIRECT AND RESIDUAL EFFECT OF GREEN MANURES AND
      OTHER ORGANIC RESIDUES ON SOIL AGGREGATION AND
      POTATO YIELD. Jour. Indian Soc. Soil Sci. 23(2):
      259-262

s7.   Stanton, W.R. 1966. GRAIN LEGUMES IN AFRICA.
      Rome: F.A.O. of the United Nations.

s8.   Sinha, S.K. 1977. FOOD LEGUMES. DISTRIBUTION,
      ADAPTABILITY AND BIOLOGY OF YIELD. Rome: F.A.O.
      Plant Production and Protection Paper #3

s9.   Skerman, P.J. 1977. TROPICAL FORAGE LEGUMES.
      Rome: F.A.O. Plant Production and Protection
      Series #2

s10. Sperbeck, Jack 1981. SCIENCE IS REDISCOVERING
      GREEN MANURES. The New Farm 3(5): 50-51

t1.   Tisdale, Samuel L. and Werner L. Nelson 1975.
      SOIL FERTILITY AND FERTILIZERS. 3rd ed. N.Y.:
      MacMillan Pub. Co. Inc.

w1.   Williams, W.A. and D.C. Finfrock 1963. VETCH
      GREEN MANURE INCREASES RICE YIELD. Ca. Ag.
      17(1): 12-13

w2.   Wilson, Harold K. 1948. GRAIN CROPS. N.Y. and
      London: McGraw-Hill Book Co. Inc.

w3.   Wolfinger, John F. 1864. GREEN MANURING AND
      MANURES. Report of the Dept. of Ag. (U.S.)

w4.   Wheeler, W.A. 1950. FORAGE AND PASTURE CROPS.
      A handbook of information about the grasses and
      legumes grown for forage in the United States.
      N.Y.: D. Van Norstrand Co. Inc.

w5.   Williams, William A. and Jean H. Dawson 1980.
      VETCH AS AN ECONOMICAL SOURCE OF NITROGEN IN
      RICE. Ca. Ag. 34(8&9): 15-16

w6.   Walters, Charles Jr. and C.J. Fenzau 1979.
      AN ACRES U.S.A. PRIMER. Raytown, Mo.: Acres U.S.A.

w7.   Wolf, Ray (ed.) Editors of Org. Gar. and Farming
      1977. ORGANIC FARMING: YESTERDAY'S AND TOMORROW'S
      AGRICULTURE. Emmaus, Pa.: Rodale Press

w8. Whyte, R.O., G. Nilsson-Leissner and H.C. Trumble
    1953. LEGUMES IN AGRICULTURE. Rome: F.A.O.
    Agricultural Studies #21
w9. Walker, N. 1975. SOIL MICROBIOLOGY. N.Y. &
    Toronto: John Wiley and Sons
y1. Yepson, Roger B. Jr. 1976. ORGANIC PLANT
    PROTECTION. Emmaus, Pa.: Rodale Press

INDEX

Mung beans 53
Mustard 54
Mycorrhizal fungi 80

Natural gas 61-62
Nectar 73
Nematode 59-60, 104
  root knot 117, 125
Nitrate 26
Nitrogen fixation 29-30,
    75
Nurse crop 76-77
Nutgrass 122

Oats 77, 79, 118
  compared to other
  species 125, 151, 154,
    171
  and intercrops 54, 174,
    176
  in rotations 123, 142,
    165
Oily seeds 85-86
Orchards 52-53, 72-73
  dryland 55
Orchardgrass 54, 112, 130-
    131
Organic acids 56-57
Oxidation 71
Oxygen 16. 20-21, 82, 86

Parasites 73
Parasitic wasps 73
Pasture 52, 88
Pea bug 152
Peaches 72
Peanuts 53, 121, 169, 178

Pearl millet 54
Peat soil 23, 64, 83, 104,
    132, 162
Pecans 73
Perennial plant 50-51, 89
pH 16, 21-22, 75, 79
Phosphate 27, 80
Phosphorus 17, 27, 52, 57,
    63, 75, 80
Physical decomposers 18
Pink bean 150
Pinto bean 150
Plow 83
Pocket gophers 106
Polysaccharides 64
Potassium 17, 27-29, 53,
    57-58, 75
Potatoes 51, 54, 66, 136,
    150, 163, 165
Predators 18, 59, 73
Protein 16, 20, 24
Purple vetch 73, 176

Quackgrass 69, 122

Rainfall 71
Rape 54
Red clover 74, 116, 132,
    139, 162, 168, 177
Red Mexican bean 150
Rhizobium 29, 60, 75, 77-
    80, 84
Rice 107, 149
Roller 83
Romans 49
Root rot 134
Rotation 48-49, 68, 73

Row crops 89
Rutabagas 109
Rye 55, 63, 77, 79, 86,
     107-108, 110, 136,
     176

Sainfoin 52, 74, 106
Saline soil 101
Sandy soil 51, 58, 76, 83,
     86, 88, 91
Sandy loam 89, 92
Sawdust 66
Scarification 80
Sclerotinia 139, 161, 166
Seed production 90
Sesbania 78, 116
Silt 58
Sludge 66
Small blue lupine 136
Smother crops 69
Sod seeder 76
Sodium 28, 67
Soil ammendments 75-76
Soil aggregates 64-65
Soil analysis 74
Soil formation 48
Soil solution 25-26, 68
Soil structure 27-28, 63-
     64
Soil temperature 22
Soil water-holding-capacity
     82
Sorghum 53, 118, 155-157
Sour clover 55, 73
Sowbugs 18
Soybeans 54, 63, 78, 147,
     149, 163, 165, 177-178

Spiders 18
Springtails 18
Spurry 51
Squash bug 73
Starch 20, 24
Starchy seeds 85-86
Strawberry clover 54
Subsoil 27, 59, 63, 83
Succession 24, 85
Sudangrass 54
Sugar beets 68, 110
Sugars 16, 20, 24, 29
Sulfur 25-26, 56, 66
     75
Sunflowers 79, 85, 173
Sweet clover 52-53, 63,
     78, 80, 112, 119, 141

Tangier pea 73
Tepary beans 54
Tetany 28
Tillage 69
Timothy 103-104, 111-112,
     119, 130, 162, 166-167
Tomatoes 55
Topsoil 27, 53, 66, 83
Toxic substances 67-68,
     72
Turnips 54, 109

Vanadium 75
Velvet beans 78, 118, 120,
     178
Vetch 63, 83, 174, 176
Vineyard 52-53
Volunteers 81